Science Projects in
ECOLOGY

Science Projects in ECOLOGY

SEYMOUR SIMON

illustrated by
CHARLES JAKUBOWSKI

Holiday House · New York

Text copyright © 1972 by Seymour Simon
Illustrations copyright © 1972 by Holiday House, Inc.
All rights reserved
Printed in the United States of America
Library of Congress Catalog card number: 76-179101
ISBN 0-8234-0204-5

For Robert

Contents

Introduction......9

The Life of a Forest......11
A Woodland Community Terrarium......17
Life in a Rotting Log......23
A Desert Community Terrarium......29
The Life of a Grassy Field......35
A Grassy Field Community Terrarium......41
The Life of a Pond......45
A Pond Community Terrarium......51
A Bog Community Terrarium......57
Constructing a Food Web......63
Studying a Plant Community......67
Plants Compete for Survival......73
A Weed's Way......77
Plant Growth and the Environment......83
Life in the Soil......89
Changing Communities in a Drop of Water......95
Chemical Cycles in Nature......101
Animal Behavior and Environment......107
Animal Structure and Environment......113
Animals Survive in Different Ways......119

Books for Research......123
Index......125

Introduction

Ecology is the study of living things and their relationships to each other and to their environment. The word ecology comes from the Greek and means a house or home. So you can think of ecology as the science that deals with the home conditions of living things.

The study of ecology takes in many different areas of nature. Ecologists want to know about the life of an individual plant or animal and groups of individuals of that species. They will study a group or community of living things in a particular area such as a pond or a rotting log. They are also interested in the communities and the environment that make up a larger area such as a redwood forest. Finally, they are interested in the sum of all the life on the earth, the biosphere.

You can study ecology on as small or as large a scale

as you like. You can make your observations out of doors on field trips in the country or in the city, or set up a miniature living community in a jar to observe in your own home. Often, one area that you study will lead you to others.

You don't need a lot of materials to do many of the projects in this book. A pencil, a notebook, and your own curiosity and interest will get you off to a good start. Other projects call for empty glass jars or similar easily obtained supplies. Some of the projects are easy to do. Other projects are more difficult. Some take only a short time to complete. Others may hold your interest for weeks or even months. Read through the book and choose a few that you would like to try. Even if you don't try them, you'll enjoy reading about the surprising and wonderful world of living things.

The Life of a Forest

Materials you will need: A pencil and a notebook, field guides to plants and animals, a small shovel, a magnifying lens, and plastic bags or jars for collecting.

Project pointers: There are many different kinds of forests, from small clumps of trees in a city park to the dense forests of the countryside. Each forest has its own kind of life. In some forests, conifers such as pine and spruce trees are the major plants, while in other forests deciduous trees such as oaks or beech or maple are more common. Animals too, differ from one forest to another.

Start your explorations on the forest floor. Feel the spongy softness of the soil. At first glance, you may not see much life on the ground. But there are more living things here than in any other level of the forest. Here is where most plant and animal life begins and here is where it all ends.

Some foresters say that each year about two tons of plant and animal materials fall on an acre of forest ground. Leaves, seeds, branches, twigs, dead animals, and dead trees all become part of the constant rain of once-living material. When you pick up some of the litter on top of the soil, you will see only the most recent additions to the forest floor, because long-dead things are buried deeper. Close to the surface, you can observe the decay processes beginning to work. Each bit of leaf and twig, each plant and animal, is broken down by soil-dwellers from earthworms to bacteria.

Dig down into the soil. The softer materials are the first to decay, but everything decays given enough time. As you dig deeper, the decaying materials become part of the soil itself. Pick up a handful of rich humus from the forest floor. In your hand are billions of living things from dozens of tiny mites and springtails to billions of microscopic bacteria. Spread out a handful of soil on a flat surface and examine it carefully for the larger of these living things. Use a magnifying lens to help you. Keep records of what you see.

Collect some of the things you find for later study at home. Use a plastic bag to collect the soil. Use small jars to collect the larger soil animals. Keep them with a bit of litter and cover the container to prevent them from drying out.

Look for millipedes, centipedes, earthworms, beetles, spiders, and other small animals. How do these animals make the soil become spongy? How does this help the soil? Look for holes and tunnels of all kinds.

The forest floor is a nursery for young trees. Look for seeds, seedlings, saplings, and trees that have not yet reached their full height. Examine the coat of a seed. Most have tough, waterproof coverings. How long do you think it will take for these seeds to sprout? Will they be able to last the winter?

Take some of the seeds home with you and try to make them sprout in some moist forest soil. It takes a year for water to get into some seeds so that they can begin to grow. Some seeds have to pass through a cold winter in order to sprout. Can you think of any advantage such delayed-sprouting seeds have over seeds that sprout quickly?

More things to try:

1. Are the temperature and light conditions the same in all parts of a forest? Use a thermometer to check temperature in the soil, just above the soil, and several feet in the air. Check these same spots at different times of the day. Which spots keep a more constant temperature? How does this affect the living things in that spot? What are the amounts of sunlight at the top level of the forest compared to lower down? Are some parts of the forest floor in deep shade even

when the sun is shining brightly? How does this affect plant and animal life?

2. What happens when it rains in a forest? Does all the water get to the soil? During a downpour, you can almost keep dry in a dense forest. Watch how the water trickles from leaf to leaf before it reaches the soil. What happens in a light rain? Does any water get through to the ground? How does the spongy forest soil hold moisture? Squeeze a handful of this soil and compare it to the soil of a field or meadow.

3. Look for the living things in the tops of the trees. You may see birds, squirrels, and flying insects. Do these also come down to the ground or spend most of their time up high? How are these animals affected by the life lower down?

4. Compare lists of animals and plants you find in different forests. Are any kinds of animals usually found in pine forests and also in oak forests? Which animals are found only in one kind or another?

A Woodland Community Terrarium

Materials you will need: An aquarium tank or a plastic shoe box, a glass cover, coarse gravel, sand, potting soil, and woodland plants and animals (see below).

Project pointers: Building a terrarium is a good way to bring a community indoors and study how its animals and plants live together. You can use almost any size container for your terrarium, even a one-gallon wide-mouthed jar. But a large container is as easy to care for as a small one, and you can set it up with a greater range of living things.

First, wash out the container carefully and rinse it clean. To grow well, plants need air around their roots and a place for excess water to drain away. Provide good

drainage with at least a one-inch-deep base of coarse gravel or small pebbles. Mixing a small amount of activated charcoal with the base material may help to keep the soil from turning sour. Slope the base material so that you have one or two rises in the back and a valley in the front. This will serve to make your landscape more attractive and to prevent puddles from forming.

Mix one part of clean sand with three parts of potting soil. Moisten the soil so that it is damp but not soggy. Spread the soil two or three inches deep over the base material, keeping the same slope. Add a small pond in one corner of your terrarium by sinking a small glass dish level with the soil and filling it with water.

Collect many different kinds of plants for a woodland terrarium. Remember to collect only what you plan to use. A few small plants look better than one large plant. Look for mosses, small ferns, lichens, and liverworts. Evergreen plants such as partridgeberry or wintergreen, and small seedlings of pine, spruce, or hemlock are interesting. You can also add a lichen-covered rock or a small branch. Keep the plants moist while bringing them home. Plastic bags are good for this. Plant the plants promptly and water them lightly around their roots.

Be sure that the glass cover is not too tight-fitting so that it allows air to circulate in the terrarium. If water droplets accumulate on the inside, provide a larger opening beneath the glass cover. It's a good idea anyway to remove the cover for a few hours every week to allow good air circulation. Remove dead leaves and decaying plants immediately. Water when the soil seems to be drying out.

You can use animals such as small frogs, toads, and salamanders in a woodland terrarium. Larger animals will uproot the plants and are not desirable. Remember that you need to feed any animals that you keep. Earthworms and small insects are good foods. Use one of the books on page 123 to check other kinds of food the animals will eat.

Keeping a terrarium lets you find out which conditions are best for a particular community. For example, how much watering is good for the plants? What happens if the humidity in the terrarium is too high or too low? Which plants grow well with high humidity; which plants with low humidity? Which seem to need different conditions for best growth? Keep records of what you find out so that you can compare your results with other sources.

More things to try:

1. Keep your terrarium near a window so that it gets some sunlight each day. Move the terrarium near a window with northern exposure for several weeks and then one with southern exposure. Which one seems to be best for plant growth? Try turning the terrarium around so that light comes from a different direction. How does this affect the plants?

2. Does one kind of a plant seem to grow quickly

and take over after a while? Can you change light or moisture conditions so that it will be held back? Try introducing another kind of plant as a possible source of competition.

3. Flowering plants such as violets and wild strawberries are difficult to grow in a terrarium, but perhaps you can try them. Try any small plant (except poison ivy) growing on a forest floor that looks interesting to you. Identify it with a field guide and keep notes on how well it does in the terrarium.

4. What happens if you increase the number of small frogs or toads in a terrarium? Is there any change in the behavior of an animal when it's crowded together with others of its kind? Do the large animals prevent the smaller ones from getting enough food? Do frogs and toads get along together? Do they both stay in the same spots, or does one species prefer a wetter or sunnier location?

Life in a Rotting Log

Materials you will need: A pencil and a notebook, a small shovel or a trowel, plastic bags, and at home, several wide-mouthed jars or an aquarium tank and a cover of fine screening.

Project pointers: A rotting log is far from being dead. In fact, a rotting log is full of life, all kinds of life. What kinds of living things you'll find depends upon how long the log has been rotting, what kind of tree it was, its location, and the time of year. But even after a tree dies and falls to the ground, it is host to a large community of living things.

Find a rotting log and look it over carefully. If it is hollow, look inside. Poke a stick in to see if anything

23

comes out. You may find some larger animals such as mice, chipmunks, a rabbit, or perhaps a snake. Small mammals often make homes inside these logs. Snakes like to hunt for food in logs because of all the living things there. Most snakes are very shy. They'll hurry away as soon as you see them. But be careful. Even though the vast majority of snakes are not poisonous, many kinds will bite if cornered and handled.

Look on the outside of the log for plants growing there. You'll often find different kinds of fungi and mosses. You may also find small seedlings of trees and wild flowers growing in decaying spots along the log.

You may find a few insects on the outside of the log, but most live inside it. Pull away a strip of loose bark. Look for dusky salamanders, small frogs, or toads. You'll probably see millipedes, centipedes, slugs, land snails, spiders, sow bugs, and beetles of all kinds. Look also for ants and termites. You may want to put some of these in small containers to examine them more carefully later.

Use your shovel to dig into the rotting wood of the log. You'll probably find passageways and tunnels of all kinds, some still in use, some left over from previous tenants. Examine the different degrees of rotting. Some parts of the log may crumble away at a touch, while other parts will still be firm. As you dig down, you'll come to the part of the log that is changing into soil. Here you'll surely find earthworms, mites, and springtails.

Take notes on all you find and what you observe.

Each log is different. Some will be newly fallen and barely decayed, while others will be mostly mounds of earth. Each will have its own life community. Each one will be a new adventure for you to explore. But don't take apart all the logs in one area. Remember that they are homes for living things. Observe what you can and come back again at a different season to see how life activities change with the passage of a year.

More things to try:

1. You can observe a rotting log community at home. Break off two or three chunks of the log with your shovel. Place the chunks in plastic bags, seal them and bring them back with you. Make a note of the conditions you found near the log such as whether it was in a damp spot and the amount of sunlight it received.

At home, place the pieces of log in wide-mouthed jars or in an aquarium tank. Spread a layer of Vaseline around the inside of the rim to keep any insects from crawling out. Cover with a fine mesh screening to allow for water evaporation.

Try to keep your rotting log terrarium under the same conditions that you found it. Water it from time to time if necessary, and keep it at about the same temperature and in the same light that you found it. Observe the kinds of insects and other animals that appear from the log. Look for tiny eggs and put them in containers to see if they hatch. Do any plants begin growing from the log? Can you see any changes taking place in the log? Each day record your finds. It might be fun to keep a daily log of a rotting log.

2. Perhaps you took home an ant colony in the piece of rotting log. Provide the ants with some food in the form of a bit of jam or a crushed insect or piece of worm. Try different foods to see which ones they seem to prefer. Do you see any ant eggs, pupas, or larvas? Do you have a very large ant? It may be the queen. Without a queen, the ant colony will not last very long.

3. What happens to the life of the log if you change some of the conditions in which you found it? For example, give the log more sunlight or more water. Do some kinds of living things seem to prosper and other kinds decline? Do such changes happen in nature?

A Desert Community Terrarium

Materials you will need: An aquarium tank or a plastic shoe box, a glass cover, coarse gravel, sand, potting soil, desert plants and animals (see below).

Project pointers: Even if you don't live near a desert you can still buy materials for a desert community near home. Clean and set up your terrarium and its base material as in the project on page 17. Mix three parts of clean sand to one part of potting soil for the sandy soil found in a desert. Spread the soil to a depth of several inches over the base material, allowing it to slope towards the front. Sink a dish level with the sand in one corner to provide water for the desert animals

Most desert plants have tough skins and small leaves

to prevent excess water evaporation and to withstand the heat. Some desert plants are annuals. These plants, mostly grasses, have seeds that sprout only after a rain and grow to maturity in a few weeks before they die. Small shrubs and lichens are also active only during the rains. Perhaps the most familiar desert plants are succulents such as cactuses. These plants often store water in their stems to tide them over during dry spells. Many have spine-like leaves that cut down the rate of water loss because they have a smaller outside surface.

Desert animals are usually small enough to hide under rocks or in burrows to escape the heat of the sun. Many are active only at night. Desert dwellers include insects and reptiles that need very little water and small mammals such as kangaroos and desert rats. Animal-eaters include foxes, small cats, skunks, and birds.

You can buy small cactus plants at many florists or garden supply stores. Get several different kinds such as a pincushion cactus, an Opuntia, a fishhook cactus, or a night-blooming cereus. You can also try different kinds of aloes and sedums. Plant these right in the sandy soil of the terrarium. Water the surface soil around the plant gently about once a week, but do not wet the rest of the terrarium. Make sure the soil is not too wet. Desert plants rot quickly in damp soil.

Animals for a desert terrarium can include a small desert tortoise, and a horned lizard (not a horned *toad*, as some pet stores mistakenly call them). Tortoises feed on fruit and small insects such as mealworms. They will also feed on succulents growing in the terrarium, so you

may have to replace these from time to time. Horned lizards should be fed a diet of ants, so be prepared with a supply.

Desert terrarium communities need lots of sunlight and do best if temperatures go up to 80° or 90° during the day. You can keep an aquarium light reflector or a 60-watt bulb over the terrarium to give it enought heat and some extra light. The glass cover should be propped open with bits of wood on all sides to allow for air circulation. If water collects on the inside of the terrarium, remove the cover until it all evaporates.

More things to try:

1. Try growing cactuses from seed. Packages of cactus seeds are sometimes available in florist shops or garden stores. Plant the seeds in the soil of the terrarium and follow the instructions on the package for their care. Examine the seeds before planting to see how they are fitted to survive the harsh environment of a desert.

2. Under certain conditions, some of the cactuses will flower. Use some of the sources listed on page 123 to find out what these conditions are and whether you can duplicate them in your terrarium.

3. Try varying amounts of sunlight and moisture to see what happens. Keep records of what you do and

what you find out. Do some kinds of plants seem to need more light or water than others? Read about these desert plants in their natural state and see where they come from and what their surroundings are like.

The Life of a Grassy Field

Materials you will need: A pencil and a notebook, field guides to help identify what you find, a small shovel, a magnifying lens, and small plastic bags or jars for collecting.

Project pointers: A grassy field is no further away from you than a back yard, a vacant lot, or a nearby park. Pick one that is convenient for you to explore, then visit it often at different times of the year. The best kind of field, in terms of the number of different animals and plants living there, is one left untended.

When you first see a weedy, overgrown field, it may not look full of life. A butterfly or two, perhaps a bird or a stray cat, may seem to be the only animals around.

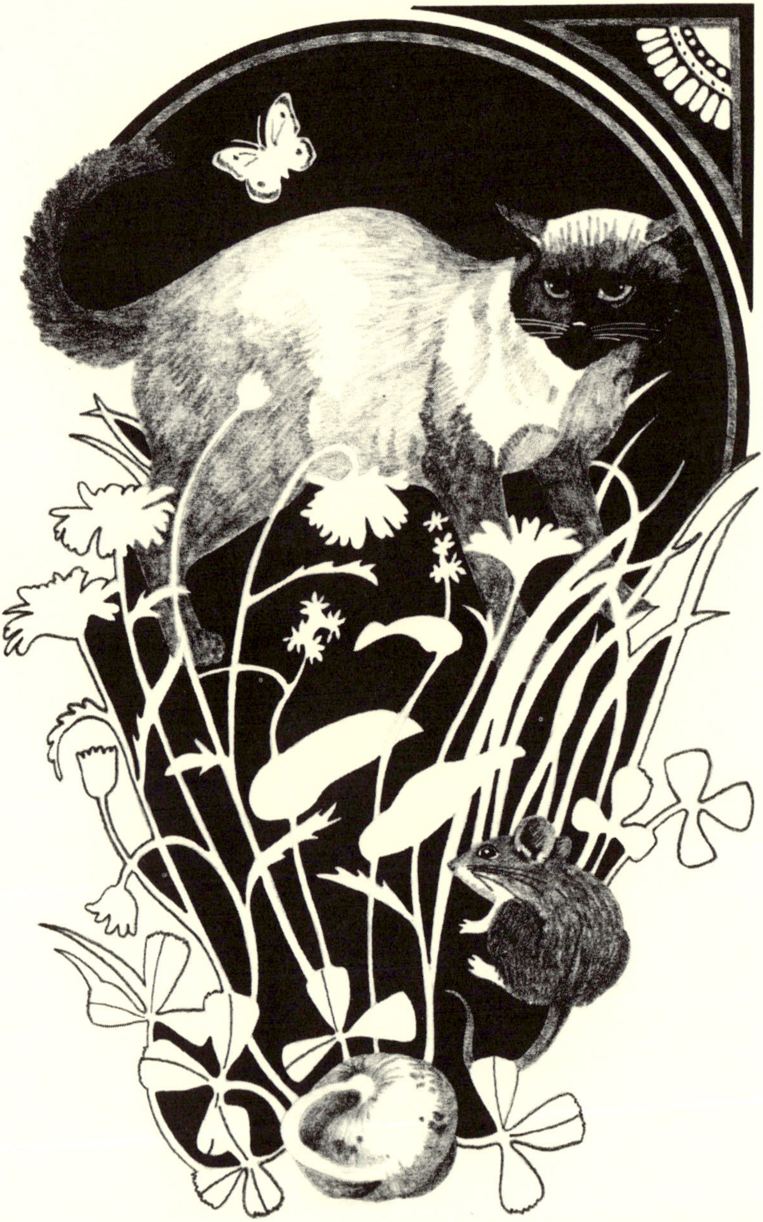

But walk into the field and look around at the ground. Look carefully. The bustling life of a field is all around you.

Perhaps the best way to start your study of a field community is to draw a rough map of the field in your notebook. Place noticeable trees or rocks in your map so that you can line it up correctly. If the field is not too large, try pacing off the edges so that you can get an idea of its size. (First, you have to measure the length of your own stride.) Draw your map to scale, such as one inch on the map equals ten feet of the field.

Now choose one section of the field to explore in detail. Note its general appearance. Does it get sun all day long, or only part of the day? Does it have a tree or some shrubs? Are there any large rocks around?

Try to identify the plants you see. Look for different kinds of grasses such as crab grass and other annuals. Larger and more showy kinds of weeds that you may find include milkweed, mullein, plaintain, ragweed, dandelion, and goldenrod. Especially look for a shiny, three-leaved vine. This is poison ivy and should not be touched or even examined at close range.

If any of these plants are in flower, examine them with a magnifying lens (but not the poison ivy). They are often beautiful to look at. Look for the developing seeds on these plants. Why do you think that one plant produces so many seeds? Do any of these plants have an odor? Crush some leaves of Queen Anne's lace and smell. Well, at least you've found out that not all plants have pleasant odors. You might like to take some of these plants home with you in plastic bags for further study.

Look through the grasses for the animal life of a field. You'll find many kinds of insects crawling among the plant stalks. One scientist found thousands of different kinds of insects in his own suburban back yard. Look for spiders of different kinds, including those that spin beautiful webs. Some of the larger animals that may live in the field are meadow mice, rabbits, woodchucks, small snakes, and birds.

Digging up some of the soil will turn up a whole new list of animals. You may find the tunnels or burrows of a mole, though you probably won't come upon a mole itself—it's too fast and has too many escape routes. You'll find earthworms, soil insects of many kinds, and jungles of plant roots. Identify what you can, record what you find, and collect those things that you would like to study later.

More things to try:

1. Compare different sections of the field in terms of the animals and plants you find there. What are the differences; what are the similarities? Which factors might account for the differences or similarities: moisture, sunlight, trees, etc?

2. Fields do not stay the same. They slowly change from bare places, to places with grasses, to places with shrubs and seedling trees, and finally to woodlands. As the plant life changes, the animal life

changes too. Which stage is your field going through? Can you see any evidence of changes taking place? Perhaps you can compare the life of other fields in different stages with the life of the field you are studying.

3. Come back to the field during different seasons of the year. How do the animals and plants change as the seasons go by? Do you find any animals during one season that don't seem to be there during any other season? What could be the reason for this? Keep records of what you find so that you won't forget.

A Grassy Field Community Terrarium

Materials you will need: An aquarium tank or a plastic shoe box, a fine screen cover, Vaseline, coarse gravel, sand, potting soil, plants and animals collected from a field.

Project pointers: Clean your container and set up the base material in it as described in the project on page 17. Mix one part of sand with three parts of potting soil. Spread the soil to a depth of several inches over the base material, following the slopes.

Gather a number of small plants from the field. Include some of the soil along with their roots. Plant them to the same depth that they were growing and water their root systems. Choose several grasses as well as small seedlings of larger plants. A rock or two may make the terrarium look more natural.

You'll probably bring in some insects along with the plants you collect, and also add any insects you find that are interesting to you. The fine screen cover and Vaseline should keep them from getting out of the terrarium. An orb-weaving spider at work is very interesting to watch. Try to catch one and include it with the animals in your terrarium. Be sure that it gets enough insects to eat. You can check its web to see if any insects are trapped. If you notice that the web has become messy, every so often clean it out so the spider has room to construct a clean new one.

Keep the terrarium in a place where it receives several hours of sunlight each day. Water every few days so that it does not dry out. Water with a fine spray so that the soil is not washed away by the heavy stream.

More things to try:

1. Look for cocoons of insect eggs on plants. Place them in the terrarium and observe them each day. What hatches from them? Insect eggs come in a bewildering variety of colors and shapes. It is very difficult to identify them. Keep records of what the eggs look like and the insect that hatches from them. Use a field guide to help you identify the insect.

2. If you want to keep a larger animal in the terrarium try to catch a small garter snake. Sink a glass dish level with the soil in one corner of the terrarium.

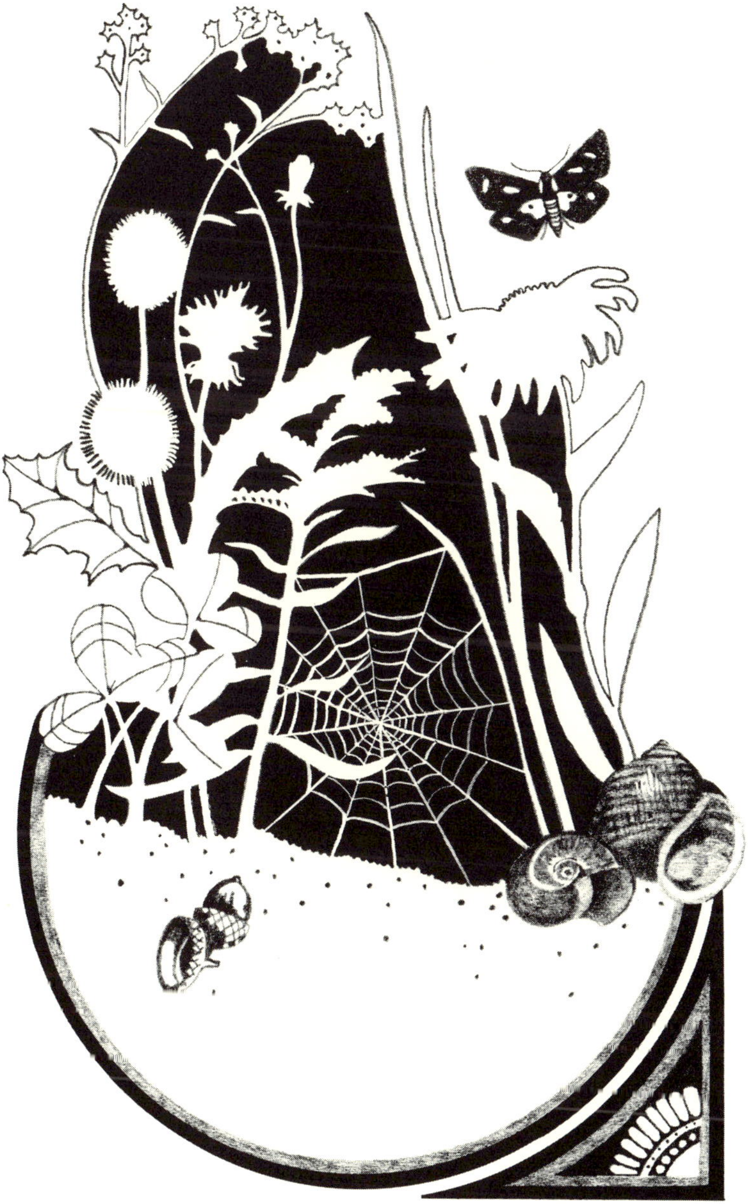

Fill it with water so that the snake can drink. For food, try earthworms or large insects. Garter snakes will also eat small frogs, toads, lizards, and fish. Water the terrarium less than usual because snakes develop skin diseases if kept in damp surroundings.

3. Do some plants grow better than others in the terrarium? Some plants may grow so well that they choke out others that you planted. Can you change light or water conditions so that the plants which are not thriving begin to pick up? What happens if the terrarium becomes completely overgrown? Do some of the plants do better in a crowded condition and others do better when less crowded? Check these plants in a field to see if they grow the same way there.

4. If you keep your terrarium for several months or longer, you may find that it goes through stages, much as a field changes with the seasons. Keep a record of what happens and see if the same thing is happening in the field. Perhaps you can keep the terrarium in an out-of-door spot, so that you can see what happens to it in the winter and then in the spring and summer of the next year.

The Life of a Pond

Materials you will need: A notebook and a pencil, field guides to help identify the animals and plants you find, a net, a bucket, a thermometer, a magnifying lens, and small plastic bags or jars.

Project pointers: A pond community is full of all kinds of interesting plants and animals. Wear a pair of old sneakers and shorts or a bathing suit when you set out to explore a pond in warm weather. Then you'll be able to get right into the pond and observe its life close up.

Just as with other living communities, all pond communities are not alike. Some ponds are mostly clear with only a few plants crowding the edges. Other ponds are quite overgrown with cattails and other plants and are

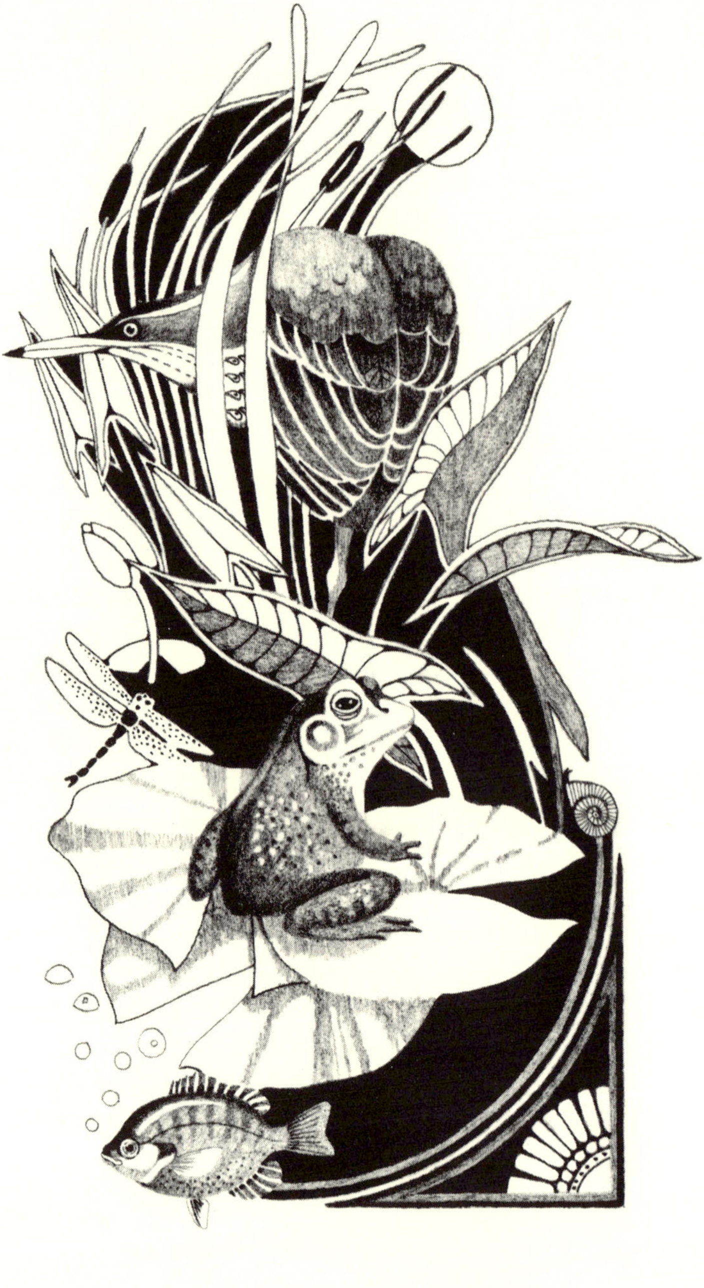

well on their way to becoming marshes or swamps. The life of a pond also varies depending upon the climate and weather of the region.

Some ponds are man-made. Some people try to construct a pond by building a dam across a stream or digging out a large hole and filling it with water. Whether natural or man-made, large or small, weedy or clear, in a city park or in the country, you'll probably be able to find a pond to explore.

Look at the different types of plants in and around a pond. On dry land beyond the boundaries of the pond, you'll find shrubs and trees. Closer to the edge of the pond, you'll find other kinds of plants growing in the wet soil. In the water but close to the shore, you'll find plants rooted in the bottom mud but emerging from the surface. Further into the pond you'll find submerged plants that are rooted in the bottom but do not grow above the surface. Finally you'll find a group of floating plants that have no roots down at the bottom. There is another group of plants too small to see without a microscope. These single-celled plants are called phytoplankton. You can read about how to study them in the project on page 95.

Not all ponds have all of these different types of plants. Try to identify the plants that you find and group them with one type or another. Do you think recently formed ponds have as many types of plants as do old ponds? Why or why not?

There are many, many different kinds of plants and animals in a pond community. Each fits into the community in a particular kind of way. The green plants, both the large ones growing around the pond and the smaller ones that you can see with a microscope, produce the food that all living things need. This process of food production by green plants is called photosynthesis. Green plants use the sun's energy to combine water, the gas carbon dioxide, and minerals into foods.

Many kinds of plant-eating animals eat the stored food of the plants. Some of the plant-eating pond animals are microscopic in size. Others such as daphnia, cyclops, and insect larvas, are a bit larger. These can be examined with a magnifying lens. Still larger plant-eating animals that live in or near ponds include snails and tadpoles.

Other animals that live in or around a pond eat the plant-eating animals. These include fishes, turtles, frogs, crayfish, snakes, birds, salamanders, and many insects. Some animals eat both plants and other animals. When a living thing dies, it may be eaten by a scavenger animal that eats dead plants or animals. Other dead animals and plants will decay and return their food energy to the pond where later it will be reused by a growing green plant.

When you identify a plant or an animal, try to place it in the food web of the pond community. Use some of the sources listed on page 123 to help you determine the food needs of any animal that you're not sure about. Use the project on page 63 to help you in constructing a food web for the pond you're exploring.

More things to try:

1. Scientists called geologists who study earth changes, state that ponds have short lives compared to most things on the earth's surface. Even the longest-lived pond lasts only a few thousand years. That may seem long to you, but compared to perhaps 5,000 million years of earth history, a pond is here today and gone tomorrow.

As soon as a pond forms, it begins to disappear. Plants grow along the shore, die and settle to the bottom. Soil washes in and begins to fill the pond. Dead animals add their bodies to the fill. The pond becomes more and more shallow. The plants grow more and more in towards the center. The pond is filled in and becomes a swamp and then a bog and finally a grassy field or a forest.

In which stage is the pond that you are exploring? Look for other ponds and try to estimate whether they are young or old. Do all ponds age at the same rate? Why or why not?

2. Use a thermometer tied to a stick to determine the temperature of the water at the surface and deeper down. Is the water temperature the same all over? Does water temperature help determine the kind of plants or animals that live there? How does the water temperature change with the seasons? What effect does this have on the pond life?

3. Many kinds of insects live in and around a pond. Some insects spend only part of their lives in the water and then part in the air. The dragonfly and the mosquito are two such insects. Dragonfly nymphs and mosquito larvas, called wrigglers, are immature stages of the insects. Make a study of the different kinds of pond insects you find. Look for aerial insects, surface-dwellers, and water-dwellers. Do these insects eat the same kinds of food during all of their lives?

A Pond Community Aquarium

Materials you will need: An aquarium tank or a wide-mouthed glass jar, gravel, a fish net, a dip tube, plastic bags, water, and pond animals and plants.

Project pointers: A pond community aquarium in your own home lets you discover many things about pond life that would be difficult to find out in nature. You can observe how a tadpole changes into a frog, or how a snail lays its eggs, or how a sunfish guards its territory.

A five- to ten-gallon aquarium tank has certain advantages over a one-gallon wide-mouthed jar. The larger aquarium allows you to keep a more representative pond community than the jar. Also, the straight sides of a tank permit you to see inside without the distortion of the curved sides of a jar. However, jars are useful for keeping animals that can't be kept with others—either because they would eat others or be eaten by them.

Start by cleaning your aquarium completely. Don't use soap or any detergent. Even a trace of leftover soap will kill pond animals. Instead, use plain water and a handful of coarse salt to rub soiled spots. Rinse the aquarium completely with running water after cleaning. It's a good idea to fill the aquarium three quarters full of water and let it stand for a day. In this way you can check for leaks and dissolve any materials that may be left in the tank. The next day, discard the test water and begin to set up the tank.

Purchase some aquarium gravel at a pet store. Do not use soil found in a pond as it will cloud the water and turn foul in a short time. Beach sand is too fine and will not permit plant roots to grow well. You need enough gravel to make a one- to two-inch layer over the bottom of the tank. Before you place it in the tank, clean the gravel by placing it in a bucket or pan and running water into it until the water comes off clean. Slope the gravel from a two-inch depth in back to one inch in front.

Now choose a spot to place the tank before putting in water. Water weighs more than eight pounds per gallon. You won't be able to carry a full aquarium. Choose a spot near a window so that the tank will get some light each day. Make sure that whatever the tank is placed upon can support its heavy weight. It might

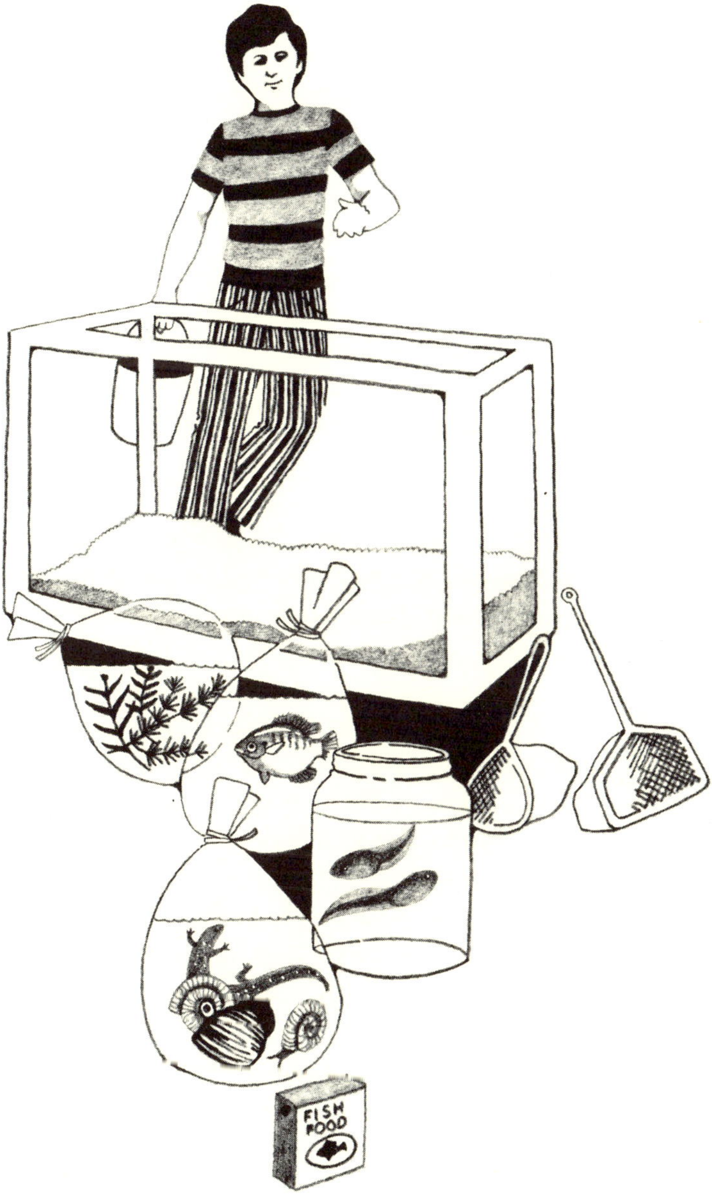
FISH FOOD

also be a good idea to place a protective sheet of oil-cloth or plastic beneath the tank.

You can use clean pond water or tap water for the aquarium. Pour the water into your cupped hand so that its force is deflected from the gravel at the bottom. Fill the tank three quarters full and rearrange any gravel which was disturbed. Allow the water to stay at least one day before adding any plants or animals to the tank. After you add the living things to the tank, fill the rest with water.

There are many different kinds of pond plants that you can use. These can either be collected in plastic bags or purchased from a pet store. Three good ones to start with are Cabomba, Vallisneria, and Anacharis or Elodea. Vallisneria should be placed in the gravel just at the level where its roots end and its leaves begin. Cabomba can be cut at one end and pushed into the gravel. It will form roots after a while. Anacharis usually doesn't form roots. You can push one end into the gravel or under a rock, or allow it to float as you like. Other plants you might put in a pond aquarium are Nitella, Lemna, and Myriophyllum.

Animals should be collected from a pond with a net and placed in glass jars or plastic bags. Try to collect smaller-sized ones—these will do better in an aquarium. Animals that you can put in an aquarium include pond snails, tadpoles, newts, crayfish, and water insects (but not mosquito larvas). Small fishes may be added as long as you don't overcrowd them. One inch of fish (not including the tail) per gallon of water is a good rule to follow. Also add some of the smaller animals such as daphnia. These are useful as food for the larger animals and are interesting in their own right. When the tadpoles develop into frogs, give them a rock or a floating platform so they can come out of the water.

Plant-eating animals such as snails and tadpoles will nibble on the plants growing in the tank, but you can add a few pieces of lettuce or spinach to supplement their diet. Animal-eaters such as fishes and newts can be fed daphnia, small worms and insects, or in the case of fishes, prepared fish foods purchased from a pet store.

More things to try:

1. If the inside walls of the aquarium become covered with green algae, you can scrape them off with an aquarium scraper. Algae are good food for the snails and tadpoles, but too many of them will make it difficult for you to see inside. Try cutting down the amount of sunlight the aquarium gets each day to control the growth of algae. You can do this by placing several sheets of tissue paper along the outside of the aquarium.

2. Some kinds of animals fight with others of their own kind. For example, sunfish will set up and defend a territory in the tank. Usually the larger ones will defend the most territory. If the size difference between two sunfish is too great, the smaller one may be prevented from ever getting enough food and will finally be hounded to death. If you see that happening to any

of your animals, return one or the other to the pond after you record the behavior.

3. Use a thermometer to check the water temperature in the tank. Do tadpoles develop faster if the temperature is high? How can you set up an experiment to test this? Do other animals behave differently when the water is cooler or hotter? In what way?

4. Do all the plants grow equally well in your aquarium, or do some do better than others? What do you think is the reason for this? Could light or temperature conditions have anything to do with it? How can you test to see what are the causes for good growth on the part of one kind of plant but not another?

5. Use a magnifying lens to examine the jelly-covered snail eggs on the glass sides of the tank. See if you can observe how the snails develop within the eggs. Try to draw what you see.

A Bog Community Terrarium

Materials you will need: An aquarium tank or a plastic shoe box, a glass cover, coarse gravel, humus, sphagnum moss, bog plants and animals.

Project pointers: A bog can be thought of as a pond on its way to becoming a woodland or field. When a pond begins to fill in with soil and decaying plants and animals, the water level begins to sink below the soil. The kinds of animals and plants that make up a pond community begin to change, and different animals and plants more suited to the water conditions of a bog take their place. The soil of a bog is quite moist, but any surface water is usually shallow, not more than a few inches deep.

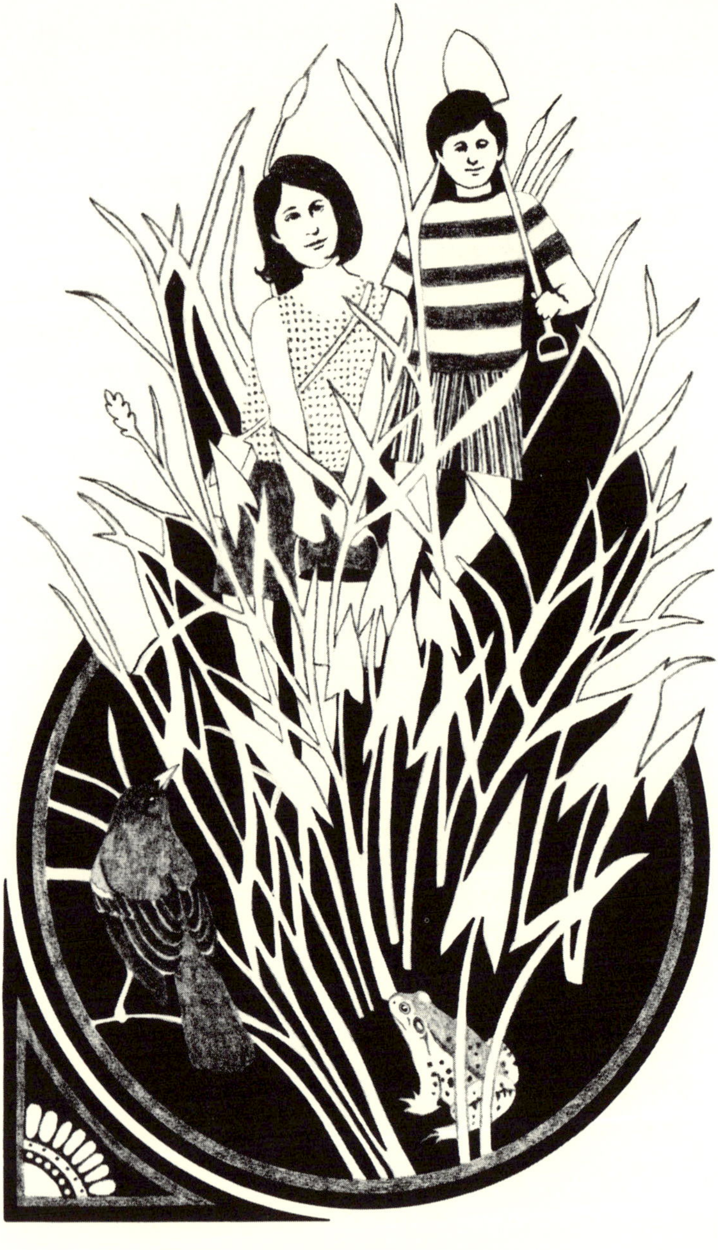

If you can, try to find a bog to explore in nature before you set up a bog terrarium at home. Take along a thermometer and check the temperature at different spots such as below the water level, on the surface, and in the air above the bog. Is there much variation? Look at some of the bog soil. Squeeze it in your hand. What is its color? How does it feel?

Note the different types of plants growing in the bog. Do you see any evidence that the kind of plant changes from the drier spots to the wetter spots of the bog? Are any of the bog plants also found in a pond or in a grassy field? Are any plants only found in bogs? Bring back a few of the smaller bog plants in plastic bags for the terrarium.

What animals or animal signs such as tracks or nests can you see around the bog? Do you see any birds living in the bog? Can you find their nests? What sort of food are they able to find? Are these birds found in other communities?

What kinds of insects do you see in the bog? Are these only found in bogs or in other places as well? You can try to trap some insects by burying an open tin can level with the surface of the bog. Bait the can with some meat or a crushed banana. What insects fall into the trap? How do these insects fit into the food web of the bog community?

Set up the bog terrarium in much the same fashion as the woodland terrarium (see page 17). After you put in the base material, mix two parts of sphagnum moss with one part of humus and spread it over the tank. Slope the surface so that you get a low spot in one corner. Adjust the water level so that some water just shows at the low spots in the tank. Check the water level every few days and add water as needed.

Perhaps the most interesting kinds of plants for a bog terrarium are insect-eating plants such as the Venus flytrap, pitcher plants, and sundews. The ends of a Venus flytrap's leaves are flattened and edged with curved spikes. The plant produces an odor which attracts insects. When an insect lands inside a leaf, it touches a hair which acts like a trigger. The leaf swiftly closes, trapping the insect inside. Then the flytrap produces a chemical which digests the insect, and allows the plant to use the minerals of its body. In a few days, the trap opens and is ready for a new victim.

A pitcher plant traps insects in a different way. The leaves form a hollow container with little hairs pointing down. Insects attracted to the odor of a pitcher plant fall into the water-filled hollow interior. The hairs prevent them from crawling out. The insect drowns and is digested. A sundew works by having sticky leaves that trap any insect that lands on them, and then digesting them.

Plant a Venus flytrap and a pitcher plant deeply in the bog soil. Sundews are shallow-rooted and are best planted close to the surface. All these plants can be purchased by mail from a scientific supply house, and may even be found in a local garden store.

Of course you can use other bog plants as well as the

insectivorous ones. Plants that you find growing in swamps or in very damp places at the edges of ponds should do well. Small salamanders, frogs, and turtles can be kept in a bog terrarium. Make sure that you feed any animals you keep.

More things to try:

1. Experiment with a Venus flytrap by putting the tip of a pencil against the guard hairs. Does the trap close quickly? How long does it take to open if there is nothing inside? Can you keep getting the leaves to shut or does it stop after a while? Catch an insect and put it in the leaf. Does the leaf stay closed now? Why do you think this is so? How can you experiment to find out?

2. Find out how much sunlight is best for the terrarium by varying the amount every few weeks. You can add to the amount of sunlight by placing an electric bulb one foot above the top of the terrarium. The best temperatures for a bog terrarium seem to be about 65° to 70° F. Do you find that to be so?

3. Do some research and find out about the different kinds of insectivorous plants in the world. Of what use are insects to a plant? Why do the insect-eating plants all grow in bogs or swamps? Can you imagine a plant large enough to eat animals such as sheep and even people? Some early naturalists thought that such

plants existed. Today, scientists don't think that there are plants like that. But you might like to read these early reports of man-eating plants. Ask a librarian to help you to track down sources for your research.

Constructing a Food Web

Materials you will need: A pencil and paper, a large sheet of oaktag or drawing paper, colored crayons, and the notes you took on the life of any community you have observed.

Project pointers: The source of food upon which all other living things depend is the green plant. A green plant is eaten by an animal, which in turn is eaten by another animal, which in turn is eaten by yet another animal, and so on. This direct line from a plant to the final animal-eater is called a food chain. It is often written in this way: Green plant→plant-eater→animal-eater 1→animal-eater 2, etc.

Try to write out some of the food chains that you

investigated in one of your community studies. One kind of pond food chain might be: plant plankton→small water insects→yellow perch→bullfrog→raccoon.

The food energy that is passed on from one link to another always shows a large amount of loss. For example, one pound of animal-eater such as a fish may need ten pounds of plant-eating insects to sustain it. Some food chains might have only three or four links; rarely are there more than five or six. Can you think why this is so?

As you construct the food chains that you find in a community, you will quickly see that an animal or plant may be part of many chains, not just one. An isolated food chain rarely exists in nature. The relationship between different food chains in a community is called a food web.

After you have written down all the food chains you can think of in the community, try combining them into a food web. Use a large poster and different colored crayons to indicate plants, plant-eaters, and animal-eaters. Use arrows to point towards the living thing using another for food. Perhaps you can use a photograph or a drawing of the plant or animal on the chart to make it more interesting to look at.

Use both direct observation and reference sources to help you in constructing your food web diagram. If you like, you can use different colored crayons for each type of information. You may need to add more large sheets of drawing paper as you make your food web more complete.

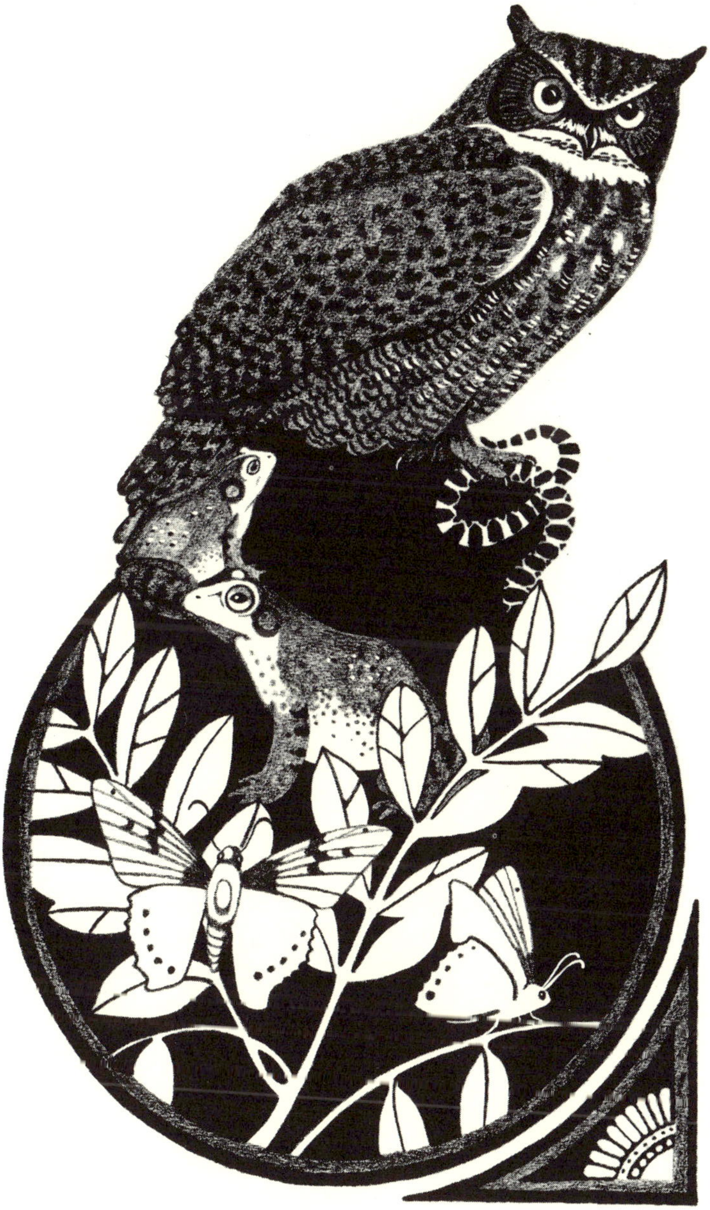

More things to try:

1. Make a food web of a community near to the one you charted. Do any animals cross from one food web to another? What are the relationships between neighboring communities? For example, does a snake eat pond animals and also animals from a nearby grassy field? Does an owl from a forest community also feed on frogs from a pond and rodents from a field?

2. How could the addition or elimination of an animal or plant from a community affect a food web? For example, what would happen to the grasses in a field community if all the foxes that ate the rabbits were eliminated? Would a new food web form?

3. What happens when man tries to rearrange a food web? In 1907, a bounty was placed on wolves, cougars, and coyotes in Arizona. The deer herd in the Kaibab Plateau in Arizona numbered about 4,000 at that time. Because of the destruction of the deer-eaters, the deer herd increased its numbers over ten times within fifteen years. By 1924, the deer herd reached 100,000. They ate all the available food, and great numbers of deer died of starvation. Man had tinkered with a food web with much different results than those he wanted. Do research to find other cases of how man affects the balance of nature. What should and should not be done when man interferes with a food web?

Studying a Plant Community

Materials you will need: A pencil and a notebook, field guides to plants, a magnifying lens, a small shovel, and plastic bags or jars for collecting.

Project pointers: Taking notes on what you find on a field trip may seem unimportant until later when you discover that you just don't remember something you observed. It's almost impossible to take too many notes; it's very easy to take too few. Jot down your findings as you make them. Take photographs or make sketches of anything that will help you remember.

First choose a plant community that you are going to study. If you are going to study it by yourself, choose a small plant community; if you are going to be

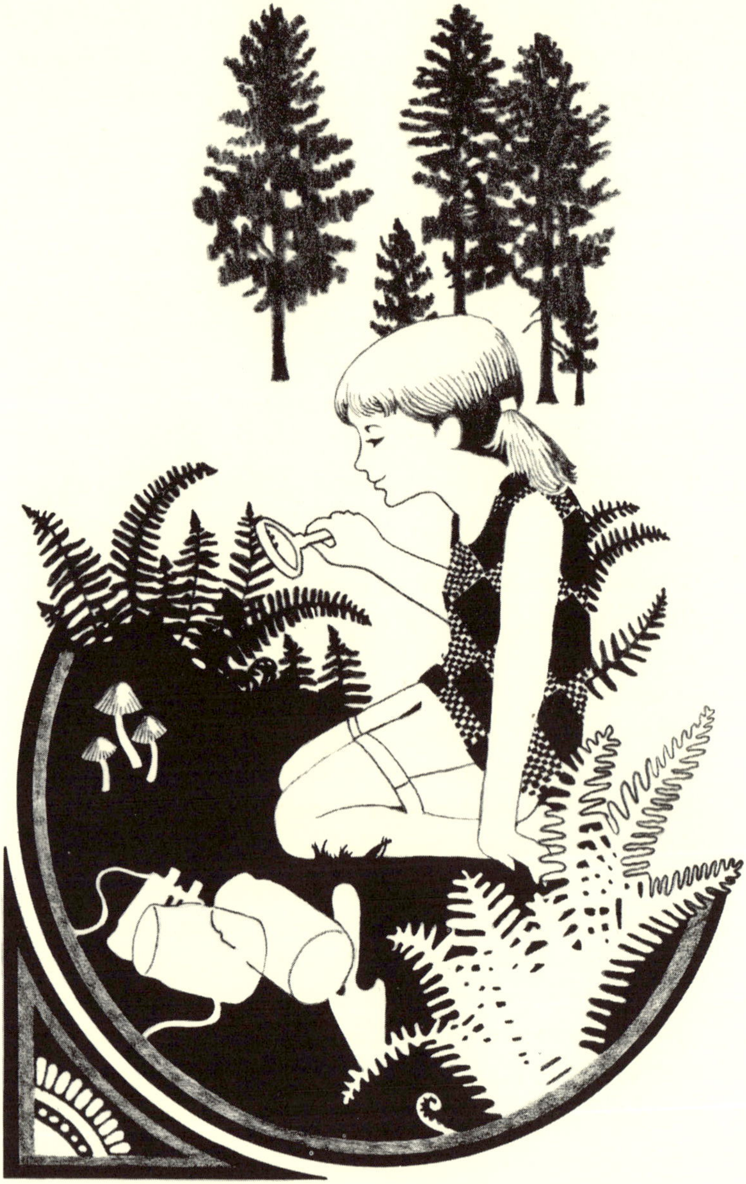

part of a group, then you can choose a larger one. The plant community you choose can be close to home or further afield. Some areas that you might study include: a clump of trees, a planting of shrubs, a lawn, a grassy field, the edge of a stream, the steep slope of a field, a vacant lot, cracks in a sidewalk.

Start out by making a plant census list. Use the plastic bags to collect one specimen of every different type of plant you find in the community you selected. Take a photograph or make a sketch of the trees or other plants too large to collect. Keep a record of the number of different kinds you collect. Identify the different plants either in the field or later if necessary with the help of a field guide.

In all communities there are certain plants that dominate. The dominant plants in a community are the ones that determine what other kinds of plants may grow around them. They do this by influencing the amount of sunlight or shade other plants get, or by their use of soil water, or by chemical means of one kind or another. Dominant plants are the largest ones present in the community in considerable numbers. For example, ragweed may be the dominant plant in a vacant lot, while oak may be the dominant plant in a clump of trees.

Nearly all plant communities have at least two layers of life. The higher layer is the one in which the dominant plant is present. But the number and kinds of plants growing in the lower layer are strongly influenced by the dominant plant. For example, tall dominant trees that block light from falling on the ground favor the kind of plant that can grow in shady areas over those that need more sunlight.

Look also for non-green plants growing on the decaying leaves and branches of the plants above. Non-green plants cannot make their own food but depend upon the materials found in other plants. The non-green plants include mushrooms, molds, and fungi of all kinds.

More things to try:

1. One kind of plant follows another kind of plant in becoming the dominant plant in a community. For example, in areas of enough rainfall the forest is the final or climax community. In areas of less rainfall, grasses may form the climax community.

Look for a forest climax community. Note that the tall trees restrict the number of shrubs or young trees growing beneath them. When these tall trees die, there will be no young trees to take their places. Then the plant community must begin all over again from the grass stage. Try to find communities in different stages and note what you think will happen to them. How long does a stage last? How long does a climax community last? What natural or man-made disturbances can affect these spans of time?

2. A forest community is called by the names of the two types of dominant trees. Thus for example, a forest made up mostly of beech and oak is called a beech-oak community. You can determine the way a forest is named in the following way. Mark off an area

in a forest of twenty-five feet by twenty-five feet. Make a count of the number of each kind of tree growing in that area. The community is named for the two trees that make up the largest percentage of trees.

3. Some kinds of trees are pioneers in an area. They grow fast but last for a relatively short time. These include such kinds as aspen, ash, hawthorn, and wild cherry. Look for woodlands composed of these trees. Are there seedlings of other kinds of trees growing up beneath them? Some seedlings of long-lived trees that you might find include maple, oak, beech, birch, hemlock, pine, and hickory. This list will change depending upon what part of the country you live in.

Plants Compete for Survival

Materials you will need: A packet of bean seeds, four milk cartons, soil, scissors, and water.

Project pointers: Every living thing needs food. Most animals spend a great deal of time searching for the food they need in large enough amounts. Since food supplies are usually limited, competition among animals is keen. Even plants that make their own food compete with each other. Each plant needs enough sunlight, enough water, and enough nutrients from the soil. Gardeners know that plants growing too close together compete for these needed materials with each other.

Use your scissors to cut open one of the long sides of each of the milk cartons. With the point of the scissors, punch one or two small holes in the opposite side for water drainage. Fill each of the cartons about three quarters full of potting soil. Water the soil in each so that it is damp but not soggy. Using your finger or the end of a pencil, make small holes ¾-inch deep and three inches apart in the first carton. Make the same size holes two inches apart in the second carton, one inch apart in the third carton, and ½-inch apart in the fourth carton.

Place a bean seed in each of the holes and cover it with soil. Place the cartons side by side in a place that gets some sunlight each day. Keep the soil in the cartons damp but not too wet. Keep a record of your observations on the growth of the seedlings in each carton. Do leaves appear on all at the same time? Do they all grow at the same rate? Do some have more leaves than others? Are some taller? Which ones look the healthiest? Do any die? What conclusions can you come to about competition among bean plants? Would this be true for all plants? How do you know?

Look at the way plants grow in nature. Can you see the effects of competition among plants growing in a field? How do young tree seedlings in a forest compete with each other? Is there any difference in the forest undergrowth in very shady spots or in clearings? How do farmers deal with plant competition in their crop plantings?

More things to try:

1. Do all seedlings need the same amount of room for good growth? Try the same experiment as above

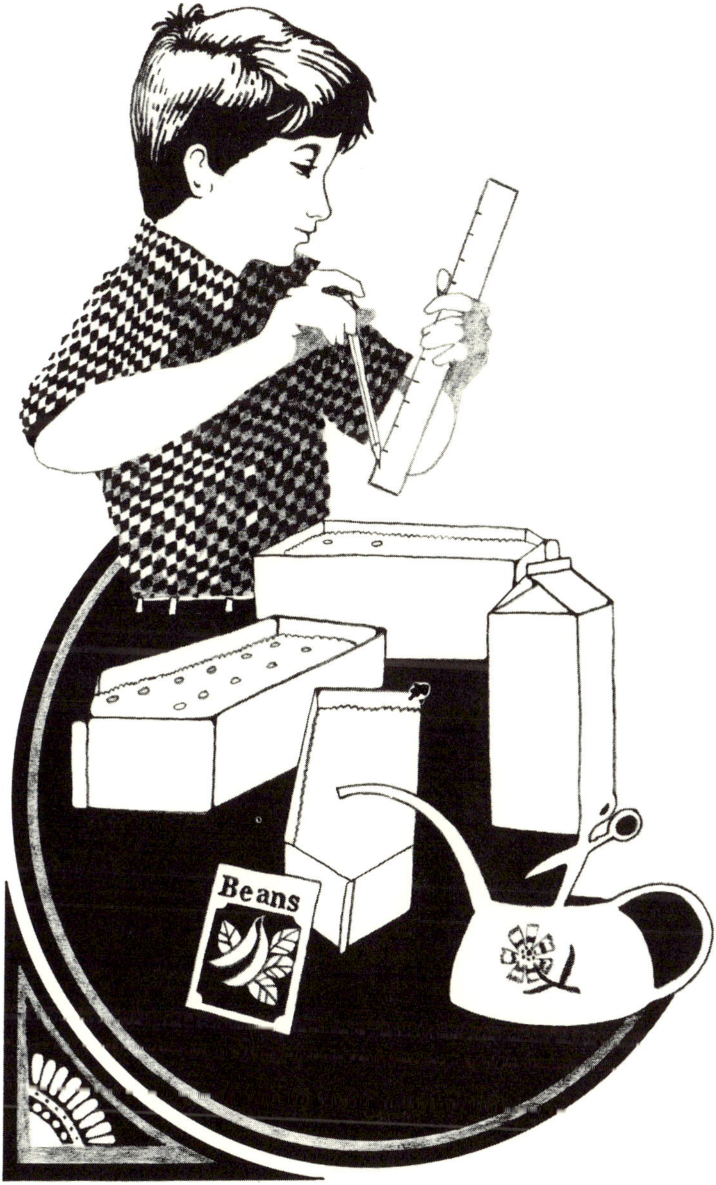
Beans

using different kinds of seeds such as carrots, corn, or dwarf marigold. Plant each seed as deeply as the instructions on the seed packet recommend, but as far apart as in the previous experiment. Compare your results with these different plants. Even if the space needs of plants are different, does the principle of competition remain the same?

2. Does using a plant fertilizer such as Hyponex change the space needs of plants? Try using the same kinds of plants with two different sets of cartons. Mix the fertilizer according to instructions on the package and use it on one of the set of cartons. Compare results. Can you see why farmers often use large amounts of fertilizers on their planting fields?

3. Would your results be any different if you kept your plant cartons in a sunnier or shadier spot? Experiment to see the effects of different amounts of sunlight on plant competition.

A Weed's Way

Materials you will need: At least five different kinds of weeds and five different kinds of plants that man uses.

Project pointers: A weed to man is just a plant growing where it is not wanted. In nature there is no such thing as a weed. So the same plant can be admired in one place and called a weed in another. Yet there are some plants that are pretty universally considered weeds. Some of the plants that we think of as weeds persist in growing in certain places no matter how hard we try to get rid of them.

Why are weeds so hard to get rid of? How are they better able to survive than most of the plants we cultivate? A plant type survives by remaining alive and

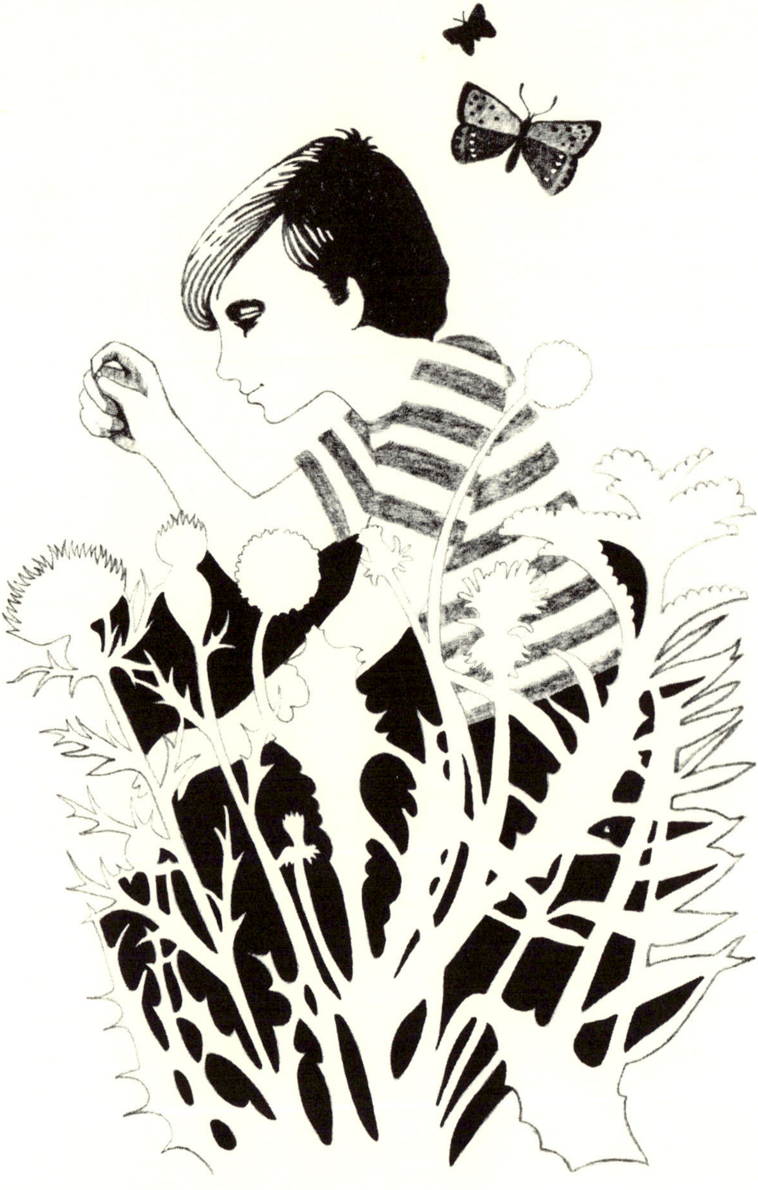

reproducing. Let's compare a weed's way of survival with a cultivated plant's way.

Select several different kinds of weeds and several kinds of wanted plants growing under the same conditions of light and moisture. Some kinds of weeds that you might select include ragweed (but not if you get hay fever), dandelion, milkweed, crab grass, thistle, mullein, plantain, or cocklebur. Select any kind of wanted plant that is available including a vegetable such as corn or tomatoes.

At home, examine each kind of leaf for a waxy coat called cutin. Cutin becomes shiny when you rub it with your finger. A heavy coat stops water loss. Pick several leaves from each of the plants and place them in separate envelopes or folded pieces of newspaper. Label each. Place the leaves in a spot where they will not be disturbed and examine them every day over a week or two. Rub the leaf between your finger to see how it is drying out. If it crumbles when rubbed gently, consider it dried out. Which plants take the longest to dry out? Why do you think this is so?

Examine the root systems of the different plants you selected. This is more easily done if you wash off all the soil. Which root systems are the largest compared to the size of the plants? Do the plants have tap roots or fibrous roots? How can a relatively heavy root system help a plant to survive?

Try to estimate the number of seeds that each plant produces. You can do this by counting carefully the number of seeds per flower and multiplying by the number of flowers. How many seeds would have to survive for a plant to maintain itself as a species? Is it likely that all plant seeds will grow? Does the number of seeds a plant produces help its survival? What conclusions can you come to about the differences in weeds compared to wanted plants?

More things to try:

1. Compare the seeds of different plants to see the ways in which they are spread. Some seeds are dispersed by air, others by water, still others by animals or mechanical means. Which kinds do you think have the best survival potential? Are weeds better in that respect?

2. Do weeds have any way of reproducing without seeds? Look for rhizomes—underground stems that grow parallel to the surface. Aerial shoots come off the rhizome at intervals producing new plants. Crab grass is a particularly good example of a plant that reproduces by rhizomes as well as by seeds. What are the advantages of this type of reproduction?

3. Are weed seeds able to last longer without sprouting than most other plants? Try planting some grass seed that is several years old. You will probably get a higher percentage of weeds sprouting than you would in new grass seed. Some weed seeds can still sprout after lying dormant for fifty years or longer.

4. Do weeds grow faster than other plants? Look at a newly planted lawn. Are the weeds growing taller and quicker than the grasses? What advantage does a tall, fast-growing plant have over smaller, slow-growing kinds? How else does a weed compete successfully with other plants?

Plant Growth and the Environment

Materials you will need: Different kinds of seeds such as bean, corn, and radish, potato eyes, a geranium or other established plant, flower pots, potting soil, sand, fertilizer, and water.

Project pointers: Plants need certain things in their environment in order to grow properly and reproduce. Without these factors being present, a plant will die or grow weakly if at all. There are many different kinds of experiments you can do to determine a plant's needs. Here are a few.

Fill a number of small flower pots with potting soil. Use two pots for each kind of seed and two pots for the potato eyes. Plant about three seeds or eyes in each pot about ½-inch deep in the soil. Keep the soil in the pots moist but not soggy.

Keep one pot containing each kind of seed that you planted near a window where it will receive good light. Keep each of the other pots in a dark place such as a closet. Keep the soil moist by watering as needed. After the seeds sprout and the plants begin to grow, compare each light plant with its opposite dark plant every day until you can come to a conclusion.

How many leaves does each plant have? Are some dark green in color and others yellowish? Is the stem thicker on one of the plants as compared to the other? Do plants grow normally in the dark? Which group of plants appears healthier?

The green material in a plant's leaf is chlorophyll. Chlorophyll is needed by a plant for photosynthesis, the process by which a plant makes food. Can plants make chlorophyll when kept away from light? Cover a leaf of a geranium or other established plant with a piece of dark paper held on with a paper clip. After a few days, remove the paper and compare the leaf with others on the plant. What differences do you observe?

To experiment with a plant's need for water in its environment, set up two pots of each plant seed as before. Start with moist soil in both. This time keep both groups of plants near a window where they will both receive good light. Water one pot of each plant seed so that the soil stays moist but not soggy. Do not water the other pots. Compare the two groups of plants each day until you can come to a conclusion. Does each

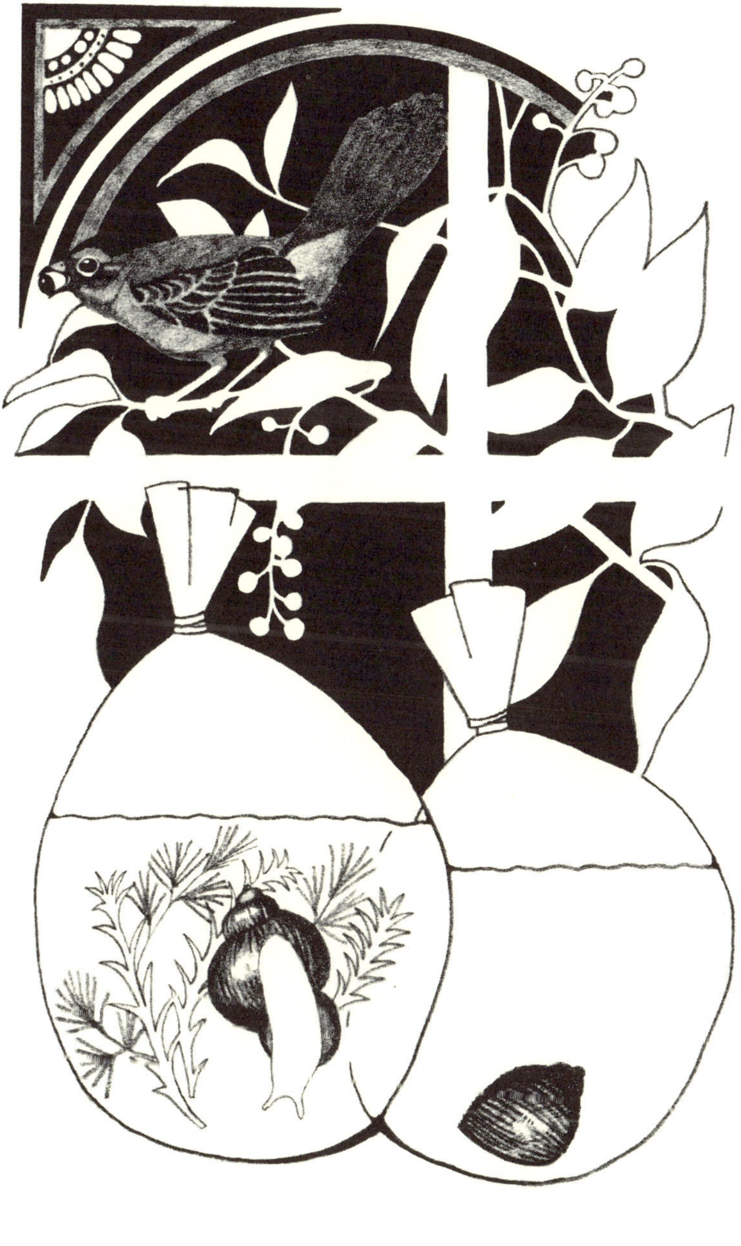

unwatered plant live about the same length of time? Do some kinds of plants need more water than others? Do all plants need water?

Green plants need certain nutrients in the soil as well as water and sunlight. Nitrogen, phosphorus, and potassium are the main minerals used by a growing plant. These mineral salts are dissolved in water and taken in by a plant through its roots.

Set up two groups of seeds in pots of sand (not potting soil) because sand does not have minerals. Keep both groups near a sunny window. Water each group every day with distilled or demineralized water, if it is available, or with tap water if necessary. As soon as leaves have developed, water one pot of each plant seed with a dissolved plant fertilizer according to directions on the package. Then continue watering all the pots with the distilled water.

Compare the growth and general appearance of the two groups for two weeks. Which group has more leaves and looks stronger? What conclusion can you come to about the need for minerals?

More things to try:

1. You can experiment with other factors in plant growth such as temperature or the amount of light or the type of soil. Use two groups for each experiment. Try to use many different plants so that your results are not just true for one particular kind. Keep daily records

of your observations and do research in books to check your results against other reported results.

2. Is there such a thing as too much water for a plant? Set up two groups of plants as before. Water one of each group so that the soil is kept fairly soggy. Keep the other one in each group moist but not soggy. Compare growth and appearance.

3. Is air needed for seeds to sprout? Put some cotton wool into two small jars. Wet the wool in each jar. Place some bean seeds on the wool. Cover one jar so that no air can enter. Leave the other jar open, but add just enough water to it to replace evaporation losses. You should get results in a few days.

Life in the Soil

Materials you will need: A shovel, plastic bags, sheets of white paper, small jars, a metal funnel, a piece of ⅛-inch metal screening to fit the funnel, a cardboard box, field guides, and a hundred watt light bulb.

Project pointers: Most soils contain large numbers of living things. In a teaspoonful of soil may be millions of bacteria, and many thousands of fungi, algae, and protozoa. Somewhat larger soil life includes animals such as springtails, worms, mites, sow bugs, millipedes, and tiny insects of all kinds.

Here's how to study the kinds of plant and animal communities that exist in soils. Using a shovel, collect a one-square-foot, two-inch deep sample of soil from a

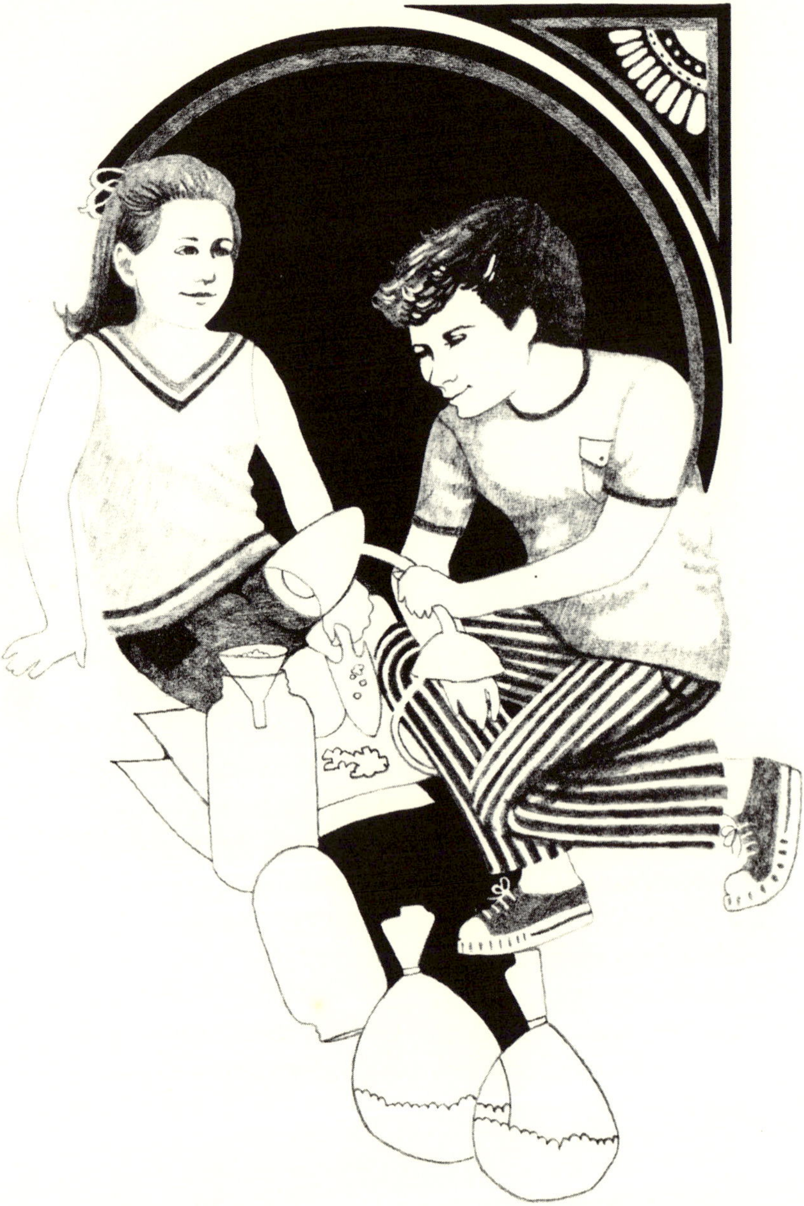

garden, a grassy field, a vacant lot, a woodland, or the bank of a pond. Pack each different soil sample separately in a plastic bag, labelling each as to where it was collected, whether it was damp or dry, and any other factors such as temperature that you may have noted.

At home examine one bag of soil at a time. Feel the soil between your fingertips and look at the color of the top and the undersurface. Is the soil very sandy? Is it loose? Is it darker on top than on bottom? Record your findings in a notebook. Spread out and examine the soil over white sheets of paper. Look for larger soil animals such as earthworms, slugs, and insect stages such as larvas and pupas. Look for small eggs and put these aside in a separate jar for possible hatching.

To separate out smaller animals from the soil sample, make a Berlese funnel as shown in the diagram. The heat from the light bulb dries out the soil, forcing the animals downward in search of moisture. The animals will fall into the collecting bottle. Use a field guide to help identify the animals that you find. Collect each type of animal in a separate small jar. Identify them and take a census of their numbers.

How many different kinds of animals did you find in your soil sample? What was the total number of animals you found? How do these compare with samples of soil taken from different places? Do you think that the soil communities will change after a period of heavy rains or of drought? Will there be a change in winter or in summer? Collect soil during such times and find out.

Use some of the sources listed on page 123 to read about the food needs of the animals you found in the soil. Can you fit these animals and plants in the soil into a food web? How does the community of living things in the soil relate to others such as a forest or grassy field community?

More things to try:

1. Take several teaspoonsful of soil and transfer each separately to a small jar containing a little water. Set this aside and examine the soil over a period of two or three weeks. Do you see any plants growing? Look for fungi of all kinds, mosses, and seedlings of larger plants.

2. Keep any eggs or immature insects in separate jars to see if they develop. Cover each jar with plastic wrap or aluminum foil to prevent any from escaping. Examine the contents of each jar every day with a magnifying lens. Keep notes and make drawings of any changes you detect.

3. If you have a microscope, use it to look for the smaller forms of life in the soil. Smear a tiny pinch of soil in a drop of water on a glass slide. Use a cover slip over it. Use a low power, then a high power objective to search for small worms, protozoa, fungi, and other forms of life. Use a microscope book to find out how to stain the slide to look for the larger kinds of bacteria.

4. Examine the decaying leaves or other plant materials you find in the soil. Which living things that you found in the soil help to change these dead plant parts into humus? What do you think would happen if these decay organisms stopped working? In nature, why is death necessary for life?

Changing Communities in a Drop of Water

Materials you will need: A microscope, clean glass slides, cover slips, a wide-mouthed glass jar and cover, a medicine dropper, water, and a handful of dead, dried grasses.

Project pointers: For this project you will need the use of a microscope. If you have never used a microscope before, make sure that you use one of the source books listed on page 123 to learn the techniques needed. Perhaps you can get permission to use a microscope in school to do your research.

Begin the project by placing several handfuls of dried grass into the wide-mouthed jar. Cover the grass with water that you have boiled for five minutes and then allowed to cool to room temperature. You can use either tap, pond, or aquarium water. Can you think of the reason why you boiled the water before placing it in the jar? Cover the jar loosely. Keep the jar in good light but not in direct sunlight. Keep a daily record of what you observe.

Starting with the first day and every day thereafter, use the medicine dropper to transfer a drop of the water in the jar to a clean glass slide. Cover the drop with a cover slip and examine it with a microscope. Use different amounts of light as you look so that any organisms will stand out clearly.

At first you may see nothing at all, but in a few days you should begin to see clumps of very tiny bacteria. Where do you think the bacteria came from? Are there large numbers of bacteria from the beginning or does the amount increase gradually? For how many days are the bacteria the only kind of organisms present?

The next kinds of living things that appear in the jar are likely to be paramecia or some other one-celled protozoans. Most will be of the same kind. Use one of the books listed on page 123 to help you to identify the organisms. Do the numbers of protozoans increase slowly or quickly? Why do you think that is so?

When do the numbers of paramecia begin to decline and other kinds of organisms begin to appear? Do you see any green algae cells? Do they increase in number and then level off? Other small organisms that you may see at this time include amoebas, rotifers, and possibly crustaceans. Try to identify the different kinds.

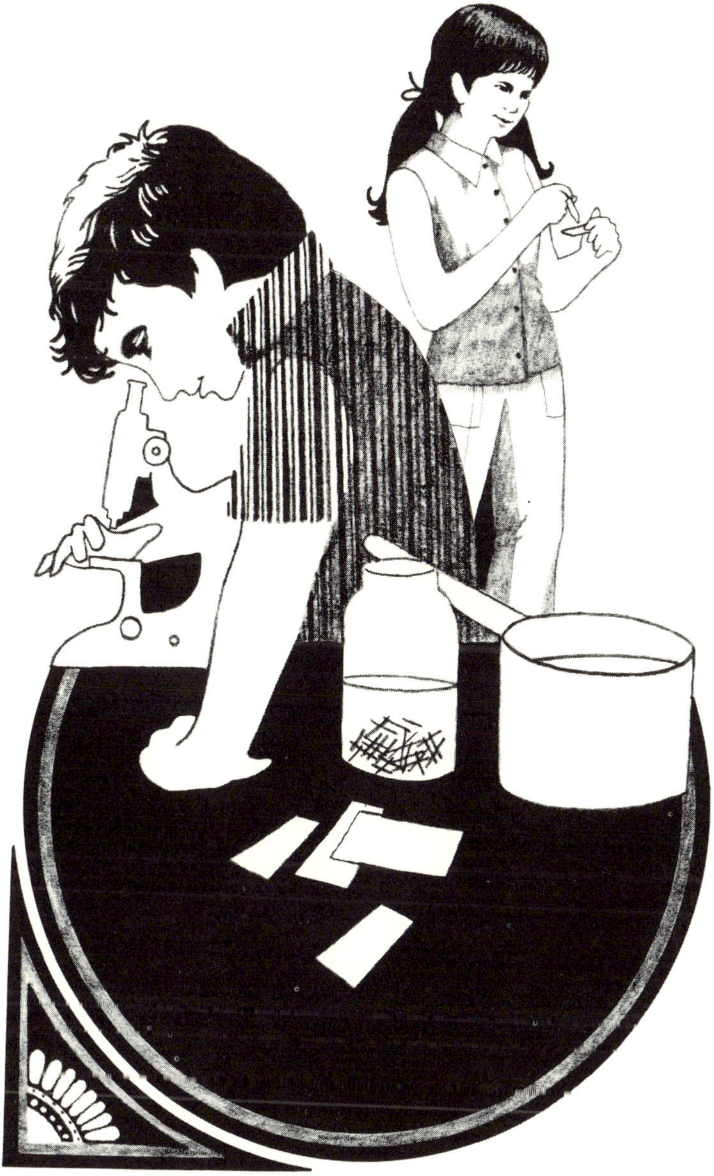

When the kinds of life in the jar seem to stay about the same for a number of days, a climax community exists. In such a community each species has enough food to support a certain number of its kind and in turn it provides other organisms with enough food to support a certain number of their kind. Communities change in a drop of water in much the same way as communities change in a larger environment such as a pond or field.

More things to try:

1. What happens if you add additional amounts of food at one of the stages before the climax community is reached? Does it prolong that stage or does it bring about the next stage more quickly? At about the time that large numbers of paramecia show up in the jar, transfer a small amount of the water into each of three clean baby-food jars. Into the first baby-food jar place a pinch of unflavored gelatin. Into the second jar place five pinches of gelatin, and into the third jar place ten pinches. Which jar becomes cloudiest? Examine a drop of water from each jar every day. Compare the number of organisms you see in the microscope field. Which has the most? Are they the same kinds or have different organisms appeared?

2. What happens if you keep the jar in direct sunlight each day? Do the same stages appear in the same order at the same times? Is the climax community the same kind? Which organisms might you expect to do

better in more light? You can also try keeping the jar in a dark place to see how the absence of light affects the communities.

3. An established aquarium should have a climax community in its water. Examine a drop of water from such an aquarium and compare it to the climax community in your jar. Is it the same or is it different? How do you account for any differences you may find?

Chemical Cycles in Nature

Materials you will need: Two plastic bags, a few aquarium or pond plants such as elodea, two pond snails, aquarium or pond water, and rubber bands.

Project pointers: Oxygen to combine with food to produce energy is needed by almost all living things. All the higher plants and animals need oxygen in its pure state, that is, not combined with any other substance. Free oxygen is also needed every time a fire burns. These two processes, respiration and combustion, are always locking up in compounds the free oxygen present in the air. In a short time, all the free oxygen would be locked up, if it weren't for green plants. Using the carbon dioxide given off by respiration and combustion,

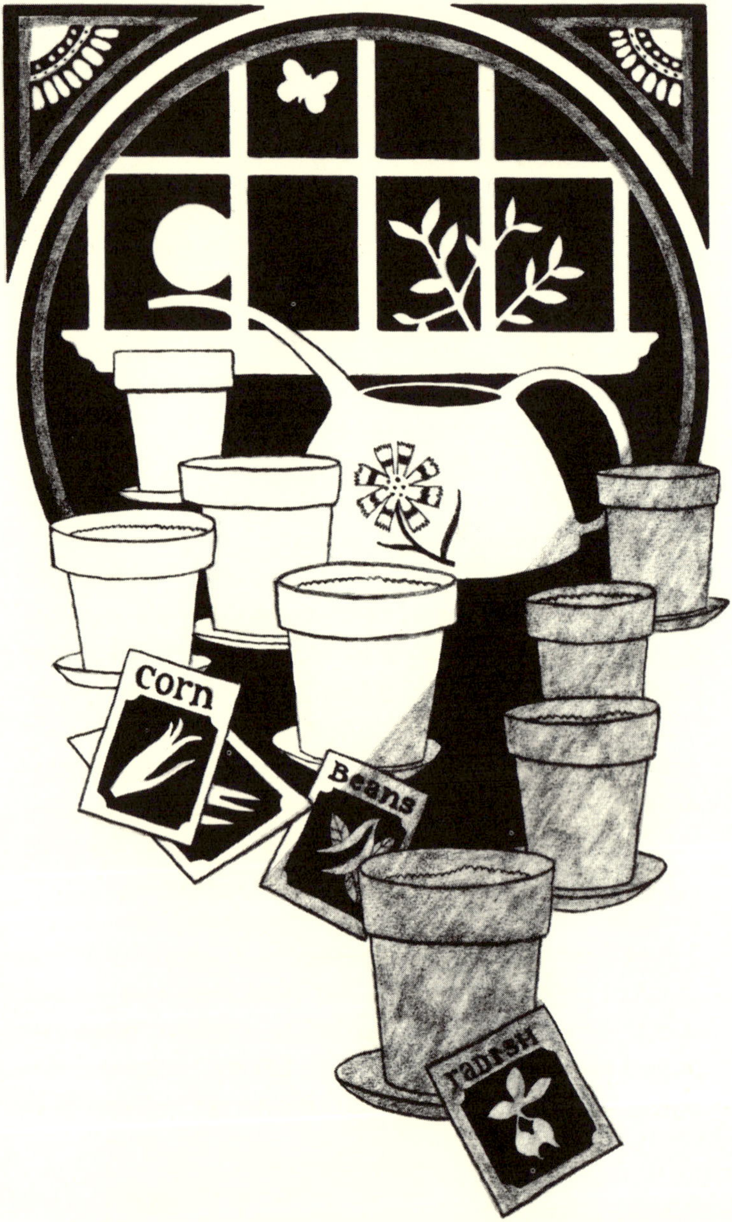
corn
Beans
radish

green plants produce large amounts of free oxygen by a process called photosynthesis.

The continual interchange of oxygen into carbon dioxide by the processes of respiration and combustion, and carbon dioxide into oxygen by photosynthesis is called the oxygen-carbon dioxide cycle. You can show how the oxygen-carbon dioxide cycle works in a simple way at home.

Half fill two plastic bags with pond or aquarium water. Place a few fresh sprigs of elodea into one of the bags. Place a pond snail into each of the bags. Blow into each bag so that it becomes full and tie the top securely with a rubber band. Place both bags in good light by a window but not in direct sunlight.

Observe what happens to the snails in each of the aquariums over a period of hours and then the next day or two. How do you explain the differences? What does the plant provide that the snail can use? What does the snail provide that the plant can use? Where does the energy that makes this chemical cycle continue come from?

More things to try:

1. Set up two plastic bags of aquarium water as before, both with elodea, but only one with a snail. Keep them in good light as before. Observe them over a period of days or even weeks. Do you see any differences in plant growth during this period of time? Why

do you think this is so? Do you know of any material other than carbon dioxide that the snail furnishes to the plant?

2. Another important cycle in nature is the nitrogen cycle. Nitrogen is part of protein, a material in living things. Nitrogen also makes up about 78% of the air, but higher plants cannot use this free nitrogen directly. Instead, plants take in soluble nitrates, containing nitrogen from the soil along with water. Animals get their nitrogen from the plants and other animals that they eat. Certain kinds of bacteria help to "fix" nitrogen from the air by combining it with hydrogen to form ammonia. This later becomes nitrate in the soil. But the main source of nitrogen for plants is living matter itself. When animals and plants die, they decay and return the nitrogen within them to the soil. Even when they are living, animals return nitrogen to the soil (or the water) with their excretions. In nature, the nitrogen cycle is more or less in balance. But man removes large amounts of nitrogen from the soil by intensive farming. How does man go about returning the nitrogen to the soil? Use books and magazines to find some ways that are used to manufacture nitrogen.

3. Check the printing on a box or bag of plant fertilizer. Some have numbers such as 5-10-5. The first number stands for the parts of nitrogen in the fertilizer, the second number for the parts of phosphorus and the third number for the parts of potassium. There are many different kinds of mixtures because different soils

must contain greater or smaller amounts of each of these chemicals for plants to grow well in them. House plant fertilizers often contain smaller amounts of chemicals called trace elements. If you have a lawn or a garden, you can investigate which of the mixtures is most suitable for your soil by applying different ones to nearby strips. Do you think that the kind of plant growing in the soil also affects the kinds of chemicals needed?

Animal Behavior and Environment

Materials you will need: Several flat glass dishes or pans, paper towels, clear food wrap, black lightproof paper, water, a light bulb, small jars for collecting, an oven timer, and four or five individuals of each of several different small invertebrate animals such as earthworms, sow bugs, or pill bugs.

Project pointers: Each different species of animal has needs which are best met in certain environments. Some animals need more moisture than others to survive, some need a warmer environment, some need darkness, and so on. In nature, each animal that you find is likely to be in a suitable environment. Those that are not either move to one that is suitable if it is available, or do not survive for very long.

Collect a few individuals of several small animals such as the ones listed at the top of the preceding page. Make a note of the conditions such as moisture, light, and temperature in which you find them. Bring them home in small jars for your studies.

For the first environmental study, place a small piece of wet towel in the center of each of the flat dishes. Place all the individuals of one kind in a separate dish and cover with clear food wrap. Leave the dishes away from any strong light or source of heat. Set the oven timer to sixty minutes (one hour) and when the bell goes off check the dishes and record where all the animals are located. Are the animals scattered throughout the dishes or are they concentrated near or away from the moisture in the center?

If there do not seem to be any differences, then try this. Use two paper towels side by side in each dish. Moisten one and leave the other dry. Place the animals in the dish and observe over a period of hours. Where do you find the animals now?

For the second study, set up the dishes with a strip of damp paper towel lining the bottom of each. Cover half of each dish with some dark, lightproof paper. Place the animals in the dishes and set them in a well lighted place. Check and record the location of the animals every fifteen minutes. Do you notice any pattern in the places that you find them? Are all the animals similar in their response to light?

For your third study, set up the containers with black lightproof paper underneath. Now place damp

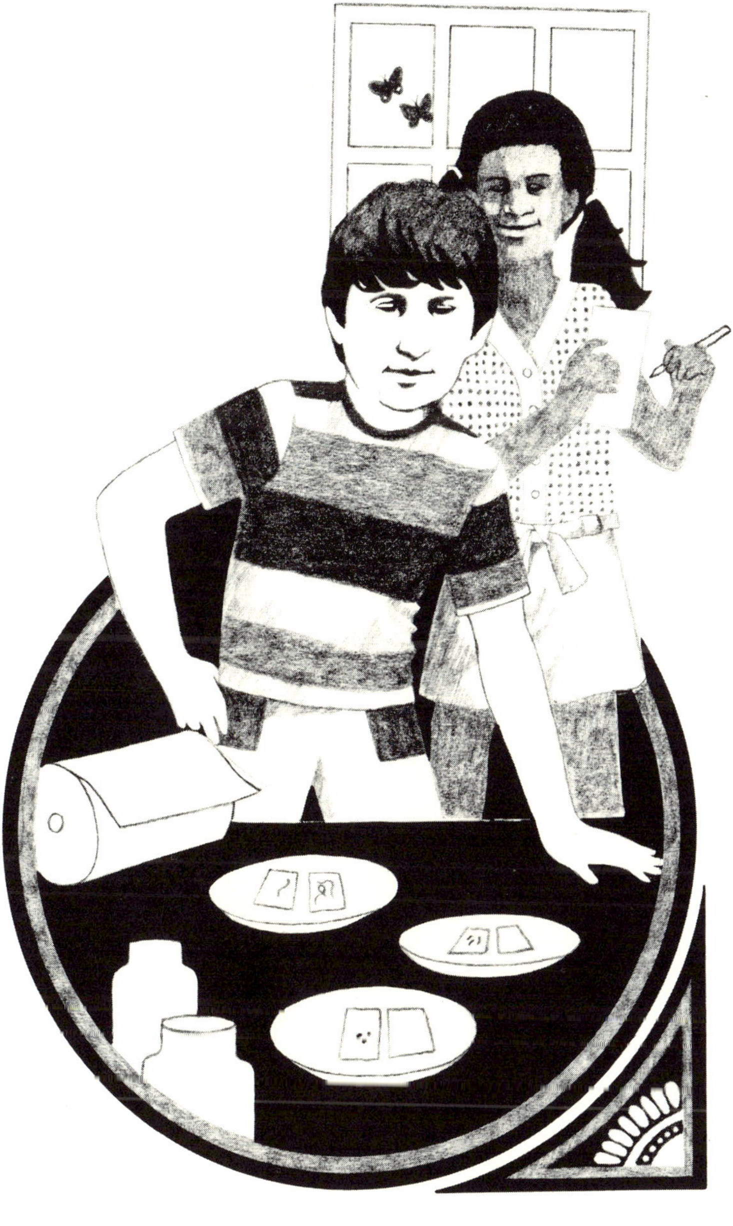

paper towel inside on the bottom. Place a light bulb under one side of each dish, so that it heats one side but doesn't give it any extra light. Place the animals in the dishes, observe, and record their locations every fifteen minutes. Make sure you stop the experiment before one side gets too hot.

How does each type of animal that you studied react to the environmental factors of moisture, light, and heat? How do these factors determine the type of environment that each animal inhabits? Check your notes about the conditions under which you found these animals in nature and see if they agree with your conclusions.

More things to try:

1. How do these animals react to other things in their environment? For example, allow the animals to move about the dish freely and then touch them with the point of a pencil. What do they do? Can you think of how pill bugs get their name? How does a reaction to touch help an animal to survive in nature?

2. Collect some small invertebrate animals from a different kind of environment and test them in the same way as you tested the animals you used. Do all animals show the same reactions? Why would you expect animals found in different places to show different environmental needs?

3. Do animals behave differently when put with other kinds of animals? Try mixing different animals together in the same dishes when you do your experiments. Are the results the same as before? What factors might influence any new behavior that you find?

Animal Structure and Environment

Materials you will need: A small amphibian such as a water-living salamander, a small lizard such as an anole, (often sold as a chameleon in a pet store) and research materials about each kind of animal.

Project pointers: A salamander and an anole look very much alike at first glance, but one spends its life on land and the other spends much of its life in water. Are there any differences in their body structures that allow these animals to adapt to their different environments? How do these differences relate to the life of the animals?

Observe the way each of the animals breathes. Look at the feathery gills just behind the head of the sala-

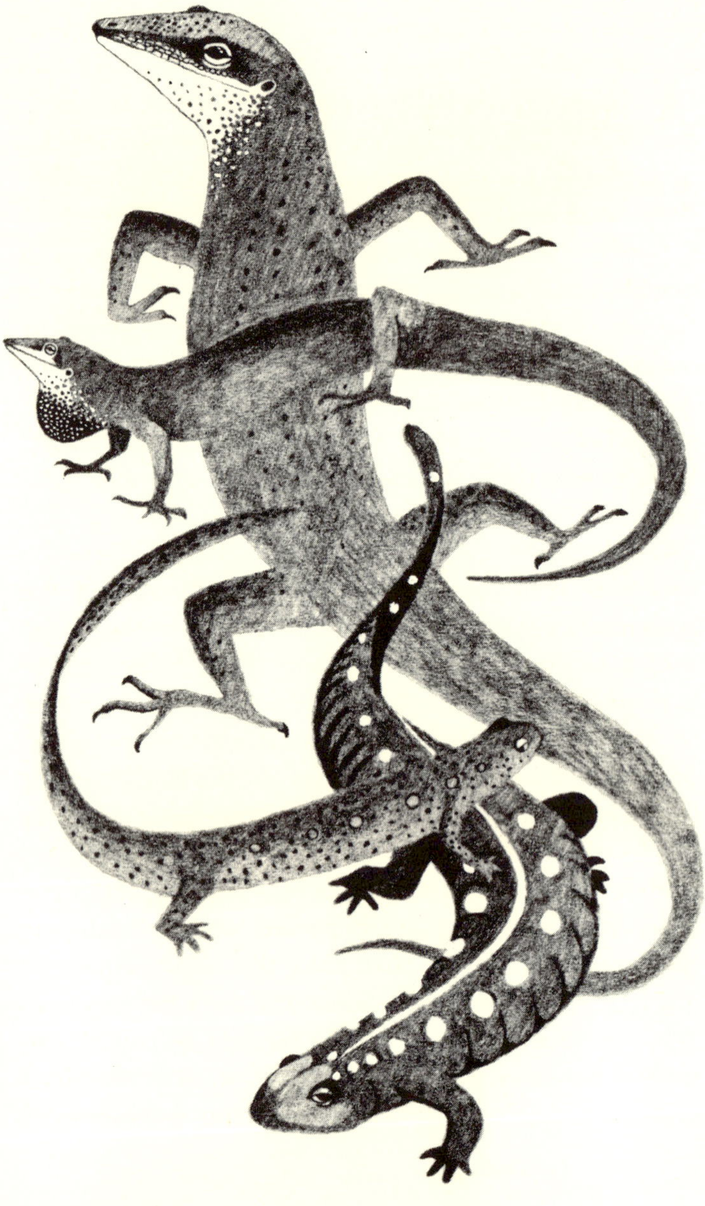

mander. (Gills are not present at all ages in all kinds of salamanders.) Do these move in a regular way? How do they assist the animal in breathing? Look at the head and neck region of the anole. Can you see any regular breathing movements? How must the anole breathe? Why do these two animals breathe differently?

Gently touch the salamander with the end of a pencil so that it swims across the container of water. In the same way, stimulate the anole to move across its container. What differences or similarities can you see in the way the animals move? How is each animal's method of moving fitted to its particular environment? How do differences in their body structures help them move?

Gently touch the skin of the salamander. Describe how it feels in your notebook. Touch the skin of the anole and describe how it feels. Do they feel the same? Which of the two animals is more likely to be able to live only in one kind of environment? Don't try this because it might result in an animal's death, but what do you think would happen if you put the land-dwelling lizard in a water environment and the water-dwelling salamander in a land environment?

Place a small earthworm or mealworm in front of the head of each of the animals. Record the way each gets the food into its mouth. Now place the earthworm about one foot away from each animal and see what happens. What are the differences in the feeding behavior? Are these differences related to the differences in the kinds of food available in a water and a land environment?

Do research to find out about the differences in the way eggs are fertilized and deposited by amphibians and by reptiles. Which kind usually deposits more eggs? How does the outer covering of the eggs of each kind differ? Is the size of an individual egg different in each species? Do the newly hatched young have the same body structure as the adult in each species?

Look at any other differences in body structure such as in the tails. Do these structures fit the animals to their environments in any way? List the differences and the similarities in body structure of the two animals and compare them.

More things to try:

1. Which of the two animals is best suited to survival at low temperatures? Try keeping each of the animals at a low temperature (but not below 40° F.) for several hours. Touch each one after this time and observe its response. Compare these responses with those of the animals when not cooled. Which is able to respond more nearly the same at low temperatures as at normal temperatures? Which animal's normal environment shows less of a temperature variation in nature?

2. Compare the body structure of the lizard and the amphibian that you studied with other body structures of a small land animal such as a hamster or a mouse and with a small water animal such as a fish. Which of these different kinds of animals would you

expect to be able to live in the largest number of different environments? Why?

3. Find out about the evolutionary history of reptiles and amphibians. How is environment related to evolution? What may happen to evolutionary history if some of the earth's environments change greatly as during an ice age?

Animals Survive in Different Ways

Materials you will need: Several friends to help you in an experiment, 250 colored toothpicks—fifty each of green, red, yellow, blue, and white.

Project pointers: Try this demonstration of how protective coloration helps an animal to survive. Choose an area of about twenty feet by twenty feet on the grassy surface of a lawn or park. Beforehand, scatter all the toothpicks throughout this area. Lead your friends to the area and inform them that they have fifteen minutes to find as many toothpicks as they can. At the end of this time collect all the toothpicks found and make a count of each color.

Which color was found least often? Why do you think this was so? Imagine your friends are birds looking for insects in the grass. Which color insect would have a better chance of survival? Is there any other kind of environment where the green toothpicks would be easily found and another color better camouflaged?

Would the same colors have the same survival value in the same place but in a different season such as winter? What might be a good color for an animal to be during a northern winter? Do you know of any animals that change colors as the seasons change?

Shape is almost as important as color in camouflage. Would green stick insects or large flat, green insects be more difficult to locate on a lawn? Which would be more difficult to locate among the leaves of an oak or maple tree? Which would be more difficult to locate in a pine or spruce tree? Why?

Different animals have many other kinds of protective coloration. Many have lighter undersides and darker topsides so that sunlight and shade produce a more even blending. Some animals have stripes or other patterns which help them blend into a particular background. Some animals are colored for reasons other than camouflage. These often have to do with attracting members of the opposite sex. Does this sound like the same reason some people dress up?

More things to try:

1. Animals sometimes cooperate with each other in order to survive. One example of this is animals that travel in groups. How does a school of fish help to protect individual fish from being eaten by a predator? What advantage do geese have when they fly in a

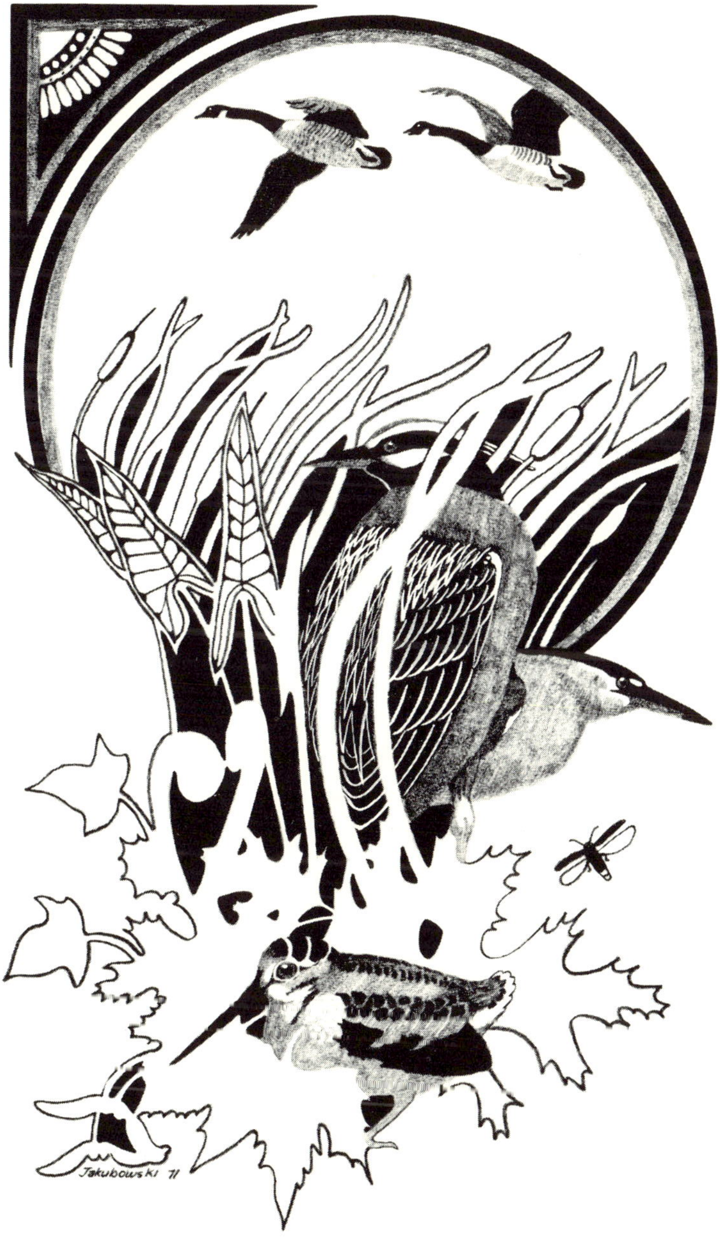
Jakubowski '71

V-formation? Try to do research and find other examples of animals cooperating for survival.

    2. Animals that live in climates that get very cold during the winter have different ways of adapting. Some animals eat very well during those seasons when food is plentiful and hibernate during cold weather. You can show hibernation in a frog at home. Place a frog in a large jar or aquarium tank with a few inches of water. Float a plastic bag full of ice cubes in the water. Observe the frog's behavior as the water temperature falls. After the frog sinks to the bottom and begins to hibernate, remove the ice and allow the frog to warm up gradually. A gradual change such as this will not hurt the frog. What other animals in your area hibernate during the winter?

    3. Another way of adapting to the cold is by migrating. Many migratory birds move north in the spring and south in the fall to escape the bad weather. Still other animals stay around during the cold weather, but grow a thick coat of fur to keep them warm. Make lists of animals in your area that migrate and those that remain. Does the severity of the weather make any difference? Do you see some animals that usually migrate or hibernate feeding in warm winter weather?

    4. Some animals (and some plants too) glow in the dark. This is called bioluminescence. Use one of the books listed on page 123 to find out in what ways these "living lights" help animals to survive. Catch some fireflies in summer and observe them at first hand.

Books for Research

Amos, W. *The Life of the Pond.* McGraw-Hill, 1967.
Beck, B. *The First Book of Weeds.* Franklin Watts, 1963.
Conklin, G. *The Bug Club Book.* Holiday House, 1966.
Collins, H. *Complete Field Guide to American Wildlife.* Harper & Row, 1959.
Farb, P. *Ecology.* Time-Life Books, 1963.
Farb, P. *The Forest.* Time-Life Books, 1963.
Farb, P. *The Insects.* Time-Life Books, 1962.
Friendly, N. *Miraculous Web: The Balance of Life.* Prentice-Hall, 1968.
Headstrom, R. *Adventures With Freshwater Animals.* J. B. Lippincott Co., 1964.
Headstrom, R. *Nature in Miniature.* Alfred A. Knopf, Inc., 1968.
Klots, E. *The New Fieldbook of Freshwater Life.* G. P. Putnam's Sons, 1966.
McCormick, J. *The Life of the Forest.* McGraw-Hill, 1966.

Milne, L. & M. *Because of a Tree.* Atheneum, 1963.
Niering, W. *The Life of the Marsh.* McGraw-Hill, 1966.
Reid, K. *Nature's Network.* Natural History Press, 1970.
Schwartz, G. *Life in a Drop of Water.* Natural History Press, 1970.
Simon, S. *Animals in Field and Laboratory.* McGraw-Hill, 1968.
Simon, S. *Exploring with a Microscope.* Random House, 1969.
Simon, S. *Science in a Vacant Lot.* Viking, 1970.
Stephen, D. & Lockie, J. *Nature's Way.* McGraw-Hill, 1969.
Villiard, P. *Reptiles as Pets.* Doubleday, 1969.
Youngpeter, J. *Winter Science Activities.* Holiday House, 1966.
Zim, H., et. al. *Golden Nature Guides.* Golden Press, various dates.

Index

algae, 55, 89, 96
ammonia, 104
amoebas, 96
amphibians, 116, 117; *see also specific amphibians*
Anacharis, 54
anoles, 113, 115
ants, 24, 27, 32
aquariums, 51, 52, 54-56, 99, 103

bacteria, 12, 89, 92, 96
beans, 74
beetles, 14, 24
bioluminescence, 123
biosphere, 9
birds, 15, 48, 50
bogs, 49, 57, 59, 61

Cabomba, 54
cactuses, 30, 32
  cereus, 30
  fishhook, 30
  Opuntia, 30
  pincushion, 30
carbon dioxide, 48, 103, 104
carrots, 76
cattails, 45
centipedes, 14, 24
chlorophyll, 84
climax community, 70, 98, 99
cocklebur, 79
combustion, 101, 103
competition, 21, 73, 74, 76
corn, 76, 79
crab grass, 37, 79, 80
crayfish, 48, 54

crustaceans, 96
cutin, 79

dandelions, 37, 79
daphnia, 48, 55
desert tortoises, 30
deserts, 29, 32
dominant plants, 69, 70
dragonflies, 50

earthworms, 12, 14, 20, 24, 38, 44, 89, 91, 92, 115
ecology, 9
Elodea, 54, 103

ferns, 19
fertilizers, 76, 104, 105
fields, 15, 35, 37-39, 44, 49, 57, 59, 66, 69, 74, 91, 92, 98
fishes, 44, 48, 54, 55; *see also specific fishes*
food chain, 63, 64
food web, 59, 64, 66, 92
forests, 9, 11, 12, 14, 15, 49, 66, 70, 74, 92
frogs, 20, 21, 24, 44, 48, 51, 55, 61, 66, 122
fungi, 24, 70, 89, 92

garter snakes, 42, 44
geraniums, 84
gills, 113
goldenrod, 37
grasses, 30, 37, 38, 41, 66, 70, 81, 95; *see also specific grasses*

hemlock, 19
hibernation, 122
horned lizards, 30
humidity, 20
humus, 12, 57, 59, 93
hydrogen, 104

larvas, 27, 91
Lemna, 54
lichens, 19, 30
liverworts, 19
lizards, 44, 115, 116; *see also specific lizards*

marshes, 47
meadows, 15
mealworms, 30, 115
migration, 122
millipedes, 14, 24, 89
milkweed, 37, 79
minerals, 48, 60, 86; *see also specific minerals*
mites, 12, 24, 89
molds, 70
moles, 38
mosquitoes, 50, 54
mosses, 19, 24, 92
mullein, 37, 79
mushrooms, 70
Myriophyllum, 54

newts, 54, 55
Nitella, 54
nitrates, 104
nitrogen, 86, 104
nitrogen cycle, 104
non-green plants, 70

orb-weaving spiders, 42
oxygen, 101, 103
oxygen-carbon dioxide cycle, 103

paramecia, 96
partridgeberry, 19
phosphorus, 86, 104
photosynthesis, 48, 84, 103
phytoplankton, 47
pill bugs, 110
pine, 19
pitcher plants, 60
plantain, 37, 79
poison ivy, 21, 37
ponds, 9, 45, 47, 49, 51, 54, 57, 59, 61, 64, 66, 91, 98
potassium, 86, 104
potato eyes, 83
protective coloration, 119, 120
protein, 104
protozoans, 89, 92, 96
pupas, 27, 91

Queen Anne's lace, 37

ragweed, 37, 69, 79
reptiles, 116, 117; *see also specific reptiles*
respiration, 101, 103
rhizomes, 80
roots, 17, 19, 41, 47, 52, 54, 60, 86
  tap, 79
  fibrous, 79
root systems, 41, 79
rotifers, 96
rotting logs, 9, 23, 24, 26, 27

salamanders, 20, 24, 48, 61, 113, 115
seeds, 14, 30, 32, 76, 79, 80, 83, 84, 86, 87
slugs, 24, 91
snails
  land, 24
  pond, 48, 51, 54, 55, 56, 103, 104
snakes, 24, 38, 42, 48, 66; *see also* garter snakes
sow bugs, 24, 89
sphagnum moss, 57, 59
spiders, 14, 24, 38; *see also* orb-weaving spiders
spruce, 19
strawberries, 21
succulents, 30; *see also* cactuses
sundews, 60
sunfish, 51, 55
swamps, 47, 49, 61

tadpoles, 48, 51, 54, 55
termites, 24
terrarium, 17, 19, 20, 21, 26, 29, 30, 32, 41, 44, 59, 61
thistle, 79
toads, 20, 21, 24, 44
tomatoes, 79
turtles, 48, 61

Vallisneria, 54
Venus flytraps, 60, 61

weeds, 77, 79, 80, 81
wintergreen, 19
wrigglers, 50